The Institute of Biology's
Studies in Biology no. 37

Photosynthesis
Third edition

D. O. Hall
Ph.D.
Professor of Biology,
King's College, University of London

K. K. Rao
Ph.D.
Honorary Lecturer,
King's College, University of London

Edward Arnold

© D. O. Hall and K. K. Rao, 1981

First published 1972
by Edward Arnold (Publishers) Limited,
41 Bedford Square, London, WC1B 3DQ

Reprinted with additions 1974
Second edition 1977
Reprinted 1978
Third edition 1981

ISBN: 0 7131 2827 5

Printed and bound in Great Britain at
The Camelot Press Ltd, Southampton

General Preface to the Series

Because it is no longer possible for one textbook to cover the whole field of biology while remaining sufficiently up to date the Institute of Biology proposed this series so that teachers and students can learn about significant developments. The enthusiastic acceptance of 'Studies in Biology' shows that the books are providing authoritative views of biological topics.

The features of the series include the attention given to methods, the selected list of books for further reading and, wherever possible, suggestions for practical work.

Readers' comments will be welcomed by the Education Officer of the Institute.

1981

Institute of Biology
41 Queen's Gate
London SW7 5HU

Preface to the Third edition

Hardly a day goes by without the importance of photosynthesis being brought to our attention. All our food and our fossil and biological fuels (biomass) are derived from the process of photosynthesis. Increasingly the products of photosynthesis are being sought to feed and fuel the world and also in the future to provide chemicals and fibres. Thus an understanding of the fundamental and applied aspects of photosynthesis is now essential to a wide range of scientists and technologists – from agriculture and forestry through ecology and biology to chemistry and engineering. It is this universality that attracts varied approaches to studying photosynthesis and makes it such an exciting field of work to so many different types of people. We hope that this becomes evident in our book.

In this third edition we have retained the inherent characteristics of the previous editions which seem to appeal to students and teachers alike, viz. basic descriptive information on the processes of photosynthesis, how this was obtained and where modern research is leading. We have updated sections on chloroplast membrane structure, electron transport and phosphorylation in plants and bacteria, C_4 and CAM photosynthesis; also the chapter on research in photosynthesis has been extended to include recent developments presented at the 1980 Photosynthesis Congress. The figures have been updated and reference list has been completely revised.

The suggestions for improvements by readers and reviewers have been most welcome. We look forward to further comments and, as we have said before, we are always open to queries and possible problem solving.

London, 1981

D.O.H.
K.K.R.

Contents

1 Importance and Role of Photosynthesis

1.1 Ultimate energy source

The term photosynthesis literally means building up or assembly by light. As used commonly, photosynthesis is the process by which plants synthesize organic compounds from inorganic raw materials in the presence of sunlight. All forms of life in this universe require energy for growth and maintenance. Algae, higher plants and certain types of bacteria capture this energy directly from the solar radiation and utilize the energy for the synthesis of essential food materials. Animals cannot use sunlight directly as a source of energy; they obtain the energy by eating plants or by eating other animals which have eaten plants. Thus the ultimate source of all metabolic energy in our planet is the sun and photosynthesis is essential for maintaining all forms of life on earth.

We use coal, natural gas, petroleum, etc. as fuels. All these fuels are decomposition products of land and marine plants or animals and the energy stored in these materials was captured from the solar radiation millions of years ago. Solar radiation is also responsible for the formation of wind and rain and hence the energy from windmills and hydro-electric power stations could also be traced back to the sun.

The major chemical pathway in photosynthesis is the conversion of carbon dioxide and water to carbohydrates and oxygen. The reaction can be represented by the equation,

$$CO_2 + H_2O \xrightarrow[\text{plants}]{\text{sunlight}} \underset{\text{Carbohydrate}}{[CH_2O]} + O_2$$

The carbohydrates formed possess more energy than the starting materials, viz. CO_2 and H_2O. By the input of the sun's energy the energy-poor compounds, CO_2 and H_2O, are converted to the energy-rich compounds, carbohydrate and O_2. The energy levels of the various reactions which lead up to the above overall equation can be expressed on an oxidation-reduction scale ('redox potential' given in volts) which tells us the energy available in any given reaction – this will be discussed later in Chapter 4. Photosynthesis can thus be regarded as a process of converting radiant energy of the sun to chemical energy of plant tissues.

1.2 The carbon dioxide cycle

The CO_2 content of the atmosphere remains almost constant in spite of its depletion during photosynthesis. All plants and animals carry out the process of respiration (in mitochondria) whereby oxygen is taken from the atmosphere by living tissues to convert carbohydrates and other tissue constituents eventually to carbon dioxide and water, with the simultaneous liberation of energy. The energy is stored in ATP (adenosine triphosphate) and is utilized for the normal functions of the organism. Respiration thus causes a decrease in the organic matter and oxygen content and an increase in the CO_2 content of the planet. Respiration by living organisms and combustion of carbonaceous fuels consume on an average about 10 000 tonnes of O_2 every second on the surface of the earth. At this rate all the oxygen of the atmosphere would have been used up in about 3000 years. Fortunately for us the loss of organic matter and atmospheric oxygen during respiration is counter-balanced by the production of carbohydrates and oxygen during photosynthesis. Under ideal conditions the rate of photosynthesis in the green parts of plants is about 30 times as much as the

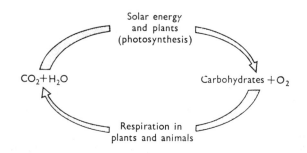

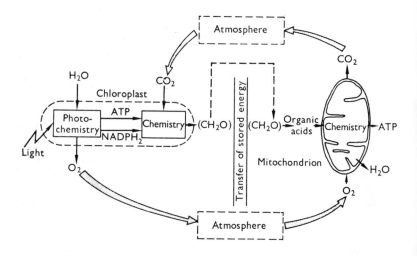

Fig. 1–1 The CO_2 and O_2 cycle in the atmosphere and the cell. (Courtesy Professor F. R. Whatley.)

rate of respiration in the same tissues. Thus photosynthesis is very important in regulating the O_2 and CO_2 content of the earth's atmosphere. The
cycle of operations can be represented as shown in Fig. 1–1. All the CO_2
in the atmosphere is cycled through plants, via photosynthesis, every 300
years and all the O_2 is cycled every 2000 years.

It should be made clear that the energy liberated during respiration is
finally dissipated from the living organism as heat and is not available for
recycling. Thus from millions of years past energy is constantly removed
from the sun and wasted as heat in the earth's atmosphere. But there is still
enough energy in the sun's atmosphere for photosynthesis to continue for
millions of years to come.

1.3   Efficiency and turnover

The solar energy striking the earth's atmosphere every year is equivalent to about 56×10^{23} joules of heat. Of this roughly half is reflected back
by the clouds and by the gases in the upper atmosphere. Of the remaining
radiation that reaches the earth's surface only 50% is in the spectral region
of light that could bring about photosynthesis, the other half being weak
infrared radiation. Thus the annual influx of energy of photosynthetically
active radiation, i.e. from violet to red light, to the earth's surface is
equivalent to about 15×10^{23} joules. However, some 40% of this is reflected by ocean surface, deserts, etc. and only the rest can be absorbed by
the plant life on land and sea. Recent estimates of the total annual amount
of biomass produced by autotrophic plants are about 2×10^{11} tonnes of
organic matter which is equivalent to about 3×10^{21} joules of energy. About
40% of this organic matter is synthesized by phytoplankton, minute plants
living near the surface of the oceans. The annual food intake by the earth's
human population (assuming the population to be 4300 million) is approximately 800 million tonnes or 13×10^{18} joules. Thus the average coefficient
of utilization of the incident photosynthetically active radiation by the
entire flora of the earth is only about 0.2% ($3 \times 10^{21}/15 \times 10^{23}$) and of this
less than 0.5% ($13 \times 10^{18}/3 \times 10^{21}$) is consumed as nutrient energy by
mankind. It is interesting that the consumption of energy by the world in
1976 was 3×10^{20} joules – this was only one-tenth of the energy stored by
photosynthesis! In fact, the energy content of the biomass standing on the
earth's surface today (90% trees) is equivalent to all our proven reserves of
fossil fuel, i.e. oil, gas and coal; also the total resources of fossil fuel stored
below the earth's surface only represents about 100 years of net photosynthesis.

1.4   Spectra

Light is a form of electromagnetic radiation. All electromagnetic radiation has wave characteristics and travels at the same speed of 3×10^8 m s^{-1}
(c, the speed of light). But the radiations differ in wavelength, the distance

between two successive peaks of the wave. Gamma rays and X-rays have very small wavelengths (less than 1000 millionth of a centimetre) while radio waves are in the order of 10^4 cm. Wave lengths of visible light are conveniently expressed by a unit called nanometre. One nanometre is 1000 millionth of a metre (1 nm = 10^{-9} m). It has been known since the time of Isaac Newton that white light can be separated into a spectrum, resembling the rainbow, by passing light through a prism. The visible portion of this spectrum ranges from the violet at about 380 nm to the red at 750 nm.

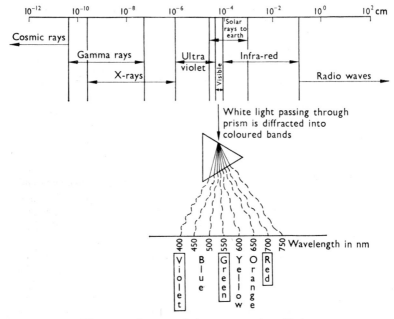

Fig. 1-2 Spectra of electromagnetic radiation.

The atmosphere of the sun consists mainly of hydrogen. The energy of the sun is derived from the fusion of four hydrogen nuclei to form a helium nucleus. $4\,{}^{1}_{1}H \rightarrow {}^{4}_{2}He + 2e^- + h\nu$ (energy). The energy liberated during the nuclear fission maintains the surface temperature of the sun around 6000 K. The sun radiates energy representing the entire electromagnetic spectrum but the earth's atmosphere is transparent only to part of the infrared and ultraviolet light and all the visible light. The ultraviolet waves which are somewhat shorter than the shortest visible light waves are absorbed by the oxygen and ozone of the upper atmosphere. This is fortunate since ultraviolet radiations are harmful to living organisms. At 6000 K, the temperature of the sun, the maximum intensity of emitted light lies in the orange part of the visible spectrum, around 600 nm.

1.5 Quantum theory

In 1900 Max Planck enunciated the theory that the transfer of radiation energy within a hot object involved discrete 'units' of energy called quanta. Planck's quantum theory can be expressed mathematically as $E = h\nu$ where E is the energy of a single quantum of radiation, ν is the frequency of the radiation (frequency is the number of waves transmitted in unit time), and h a constant. The Planck's constant (h) has the dimensions of the product of energy and time and its value in the c.g.s. system is 6.626×10^{-34} J s. Planck's theory proposes that an oscillator of fundamental frequency (ν) would take up energy $h\nu$, $2h\nu$, $3h\nu \to n h\nu$, but it could not acquire less than a whole number of energy quanta. Five years later Albert Einstein extended Planck's theory to light and proposed that light energy is transmitted not in a continuous stream but only in individual units or quanta. The energy of a single quantum of light or *photon* is the product of the frequency of light and Planck's constant, i.e. $E = h\nu$. Since frequency is inversely related to wavelength, it follows that photons of short wave light are more energetic than photons of light of longer wavelength, i.e. at one end of the spectrum photons of blue light are more energetic than those of red light at the other end.

For photosynthesis to take place the pigments present in plant tissues should absorb the energy of a photon at a characteristic wavelength and then utilize this energy to initiate a chain of photosynthetic chemical events. We will learn later that an electron is ejected from the pigment immediately after the absorption of a suitable quantum of light. It should be emphasized that a photon cannot transfer its energy to two or more electrons nor can the energy of two or more photons combine to eject an electron. Thus the photon should possess a critical energy to excite a single electron from the pigment molecule and initiate photosynthesis. This accounts for the low efficiency of infra-red radiation in plant photosynthesis since there is insufficient energy in the quantum of infra-red light. Certain bacteria, however, contains pigments which absorb infra-red radiation and carry out photosynthesis which is quite different from plant-type photosynthesis in that no O_2 is evolved during the process (see Chapter 7).

1.6 Energy units

According to Einstein's law of photochemical equivalence a single molecule will react only after it has absorbed one photon of energy ($h\nu$). Hence one mole (gram-molecule) of a compound must absorb N ($N = 6.023 \times 10^{23}$, the Avogadro number) photons of energy, i.e. $Nh\nu$, to start a reaction. The total energy of photons absorbed by one mole of a compound is called an Einstein.

Let us calculate the energy of a mole (or Einstein, i.e. 6.023×10^{23}

quanta) of red light of wavelength 650 nm (6.5×10^{-7} m). The frequency, $v = c/\lambda =$ speed of light/wavelength of light

$$v = 3.0 \times 10^8/6.5 \times 10^{-7} = 4.61 \times 10^{14}$$

$E = Nhv$, i.e. Energy = number of molecules $\times$ Planck's constant $\times$ frequency

$$\therefore\; E = 6.023 \times 10^{23} \times 6.626 \times 10^{-34} \times 4.61 \times 10^{14}$$

$$= 18.40 \times 10^4 \text{ joules} = \text{energy of one Einstein of red light}$$

$$\text{or } E = 18.40 \times 10^4/4.184 \times 10^3 = 43.98 \text{ kcal}$$

(One kilocalorie, kcal, is equal to 4.184×10^3 joules.) Thus 1 mole of red light at 650 nm contains 18.40×10^4 joules of energy.

The energy of photons can also be expressed in terms of electron volts. An electron volt, eV, is the energy acquired by an electron when it falls through a potential of 1 volt, which is equal to 1.6×10^{-19} joules. If 1 mole of a substance acquires an average energy of 1 eV the total energy of the mole (6.023×10^{23} molecules) can be calculated to be 9.64×10^4 joules. Thus the energy of 1 mole of 650 nm light is equal to 1.91 eV ($18.40 \times 10^4/9.64 \times 10^4$.

Table 1   Energy levels of visible light.

Wavelength	Colour	Joules per mole	kcal per mole	Electron volts per mole
700 nm	Red	17.10×10^4	40.87	1.77
650 nm	Orange-red	18.40×10^4	43.98	1.91
600 nm	Yellow	19.95×10^4	47.68	2.07
500 nm	Blue	23.95×10^4	57.24	2.48
400 nm	Violet	29.93×10^4	71.53	3.10

2 History and Progress of Ideas

2.1 Early discoveries

In the early half of the seventeenth century the Flemish physician van Helmont grew a willow tree in a bucket of soil feeding the soil with rain water only. He observed that after five years the tree had grown to a considerable size though the amount of soil in the bucket had not diminished significantly. Van Helmont naturally concluded that the material of the tree came from the *water* used to wet the soil. In 1727 the English botanist Stephen Hales published a book in which he observed that plants used mainly *air* as the nutrient during their growth. Between 1771 and 1777 the famous English chemist Joseph Priestley (who was one of the discoverers of oxygen) conducted a series of experiments on combustion and respiration and came to the conclusion that green plants were able to reverse the respiratory processes of animals. Priestley burnt a candle in an enclosed volume of air and showed that the resultant air could no longer support burning. A mouse kept in the residual air died. A green sprig of mint, however, continued to live in the residual air for weeks. At the end of this time Priestley found that a candle could burn in the reactivated air and a mouse could breathe in it. We now know that the burning candle used up the *oxygen* of the enclosed air which was replenished by the photosynthesis of the green mint. A few years later the Dutch physician, Jan Ingenhousz, discovered that plants evolved oxygen only in *sunlight* and also that only the *green* parts of the plant carried out this process.

Jean Senebier, a Swiss minister, confirmed the findings of Ingenhousz and observed further that plants used as nourishment *carbon dioxide* 'dissolved in water'. Early in the nineteenth century another Swiss scholar, de Saussure, studied the quantitative relationships between the CO_2 taken up by a plant and the amount of organic matter and O_2 produced and came to the conclusion that *water* was also consumed by plants during assimilation of CO_2. In 1817 two French chemists, Pelletier and Caventou, isolated the green substance in leaves and named it *chlorophyll*. Another milestone in the history of photosynthesis was the enunciation in 1845 by Robert Mayer, a German physician, that plants transform energy of sunlight into chemical *energy*. By the middle of the last century the phenomenon of photosynthesis could be represented by the relationship

$$CO_2 + H_2O + light \xrightarrow[\text{plant}]{\text{green}} O_2 + \text{organic matter} + \text{chemical energy}$$

Accurate determinations of the ratio of CO_2 consumed to O_2 evolved during photosynthesis were carried out by the French plant physiologist Boussingault. He found in 1864 that the photosynthetic ratio – the volume of O_2 evolved to the volume of CO_2 used up – is almost unity. In the same year the German botanist Sachs (who also discovered plant respiration) demonstrated the formation of *starch* grains during photosynthesis. Sachs kept some green leaves in the dark for some hours to deplete them of their starch content. He then exposed one half of a starch-depleted leaf to light and left the other half in the dark. After some time the whole leaf was exposed to iodine vapour. The illuminated portion of the leaf turned dark violet due to the formation of starch-iodine complex; the other half did not show any colour change.

The direct connection between oxygen evolution and chloroplasts of green leaves and also the correspondence between the action spectrum of photosynthesis and the absorption spectrum of chlorophyll (see Chapter 4) were demonstrated by Engelmann in 1880. He placed a filament of the

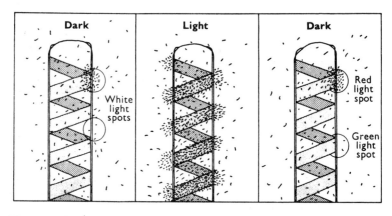

Fig. 2–1   Summary of Engelmann's experiment for studying photosynthesis using the alga *Spirogyra* and motile bacteria. The alga has a spiral chloroplast and the bacteria migrate towards regions of higher O_2 concentration. Left: Illumination with a spot of white light. Centre: Complete illumination with white light. Right: Illumination with spots of red and green light. Note the lack of O_2 evolution in green light.

green alga *Spirogyra*, with its spirally arranged chloroplasts, on a microscope slide together with a suspension of an oxygen-requiring, motile bacteria. The slide was kept in a closed chamber in the absence of air and illuminated. Motile bacteria would move towards regions of greater O_2 concentration. After a period of illumination the slide was examined under a microscope and the bacterial population counted. Engelmann found that the bacteria were concentrated around the green bands of the algal filament. In another series of experiments he illuminated the alga with a spectrum of light by interposing a prism between the light source and the microscope stage. The largest number of bacteria surrounded those parts of the algal filament that were in the blue and red regions of the spectrum. The chlorophylls present in the alga absorbed blue and red light; since light has to be absorbed to bring about photosynthesis Engelmann concluded that chlorophylls are the active photoreceptive *pigments* for photosynthesis. The state of knowledge on photosynthesis at the beginning of this century could be represented by the equation:

$$(CO_2)_n + H_2O + light \xrightarrow[\text{plant}]{\text{green}} (O_2)_n + starch + chemical\ energy$$

2.2   Further work related to techniques

Though by the beginning of this century the overall reaction of photosynthesis was known, the discipline of biochemistry had not advanced enough to understand the mechanism of reduction of carbon dioxide to carbohydrates. It should be admitted that even now we know very little about certain aspects of photosynthesis. Attempts were made early to study the effects of light intensity, temperature, carbon dioxide concentration, etc. on the overall yields of photosynthesis. Though plants of divergent species were used in these studies, most of the determinations were carried out with unicellular green algae, *Chlorella* and *Scenedesmus*, and the unicellular flagellate *Euglena*. Unicellular plants are more suitable for quantitative research since they can be grown in all laboratories under fairly standard conditions. They can be suspended uniformly in aqueous buffer solutions and aliquots of the suspension can be transferred with a pipette as though they were true solutions. Chloroplasts are best prepared and studied from leaves of higher plants, the most common being spinach leaves as they are usually available fresh in the market and can be grown quite easily; peas and lettuce are also sometimes used.

Since CO_2 is fairly soluble and O_2 is relatively insoluble in water, during photosynthesis in a closed system there will be a change in gas pressure. The Warburg respirometer (adapted by Otto Warburg in 1920) is often used in studies involving the action of light on photosynthetic systems by measuring the changes in the O_2 volume of the system (see *Manometric Techniques*, UMBREIT, BURRIS and STAUFFER, Burgess Publ. Co., U.S.A., for details).

The oxygen electrode is a more convenient instrument to measure uptake or liberation of O_2 during a reaction. The electrode works on the principle of polarography and is sensitive enough to detect O_2 concentrations of the order of 10^{-8} moles cm^{-3} (0.01 millimolar). The apparatus consists of platinum wire sealed in plastic as cathode, and an anode of circular silver wire bathed in a saturated KCl solution. The electrodes are separated from the reaction mixture by an O_2 gas-permeable teflon membrane. The reaction mixture in the plastic (or glass) container is stirred constantly with a small magnetic stirring rod. When a voltage is applied across the two electrodes, with the platinum electrode negative to the reference electrode, the oxygen in the solution undergoes electrolytic reduction. The flow of current in the system between 0.5 and 0.8 V varies in a linear relationship to the partial pressure of the oxygen in solution. The instrument is usually operated at a voltage of about 0.6 V. The current liberated is measured by connecting the electrode set up to a suitable recorder. The whole apparatus is kept at a constant temperature by circulating water from a controlled

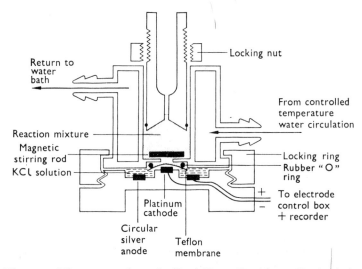

Fig. 2–2   The oxygen electrode (Rank Bros., Bottisham, Cambridge).

temperature water source. The effects of light and of various chemicals on photosynthesis are measured using the oxygen electrode. The O_2 electrode has the advantage over the Warburg method in that rapid and continuous measurements of O_2 evolution can be made. However, the Warburg apparatus can measure up to 20 reaction mixtures simultaneously while the O_2 electrode measures reactions one at a time.

2.3   Limiting factors

The extent of photosynthesis performed by a plant depends on a number of internal and external factors. The chief internal factors are the structure of the leaf and its chlorophyll content, the accumulation within the chloroplasts of the products of photosynthesis, the influence of enzymes and the presence of minute amounts of mineral constituents. The external factors are the quality and quantity of light incident on the leaves, the ambient temperature and the concentration of carbon dioxide and oxygen in the surrounding atmosphere.

2.3.1   The effect of light intensity

The effect of light intensity on the photosynthetic activity of a healthy suspension of *Chlorella* cells is illustrated in Fig. 2–3. At low light intensities the rate of photosynthesis, as measured by oxygen evolution, increases linearly in proportion to light intensity. This region of the curve, marked X, is known as the light-limiting region. With more and more light intensity, photosynthesis becomes less efficient until after about 10 000 lux

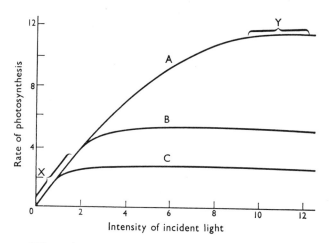

Fig. 2–3   Effect of external factors on rate of photosynthesis. A, Effect of light intensity at 25 °C and 0.4% CO_2; B, at 15 °C and 0.4% CO_2; C, at 25 °C and 0.01% CO_2. All units in the graph are arbitrary.

(1000 foot candles) increasing light intensity produces no further effect on the rate of photosynthesis. This is indicated by the horizontal parts of the curves in the figure. This plateau region designated Y is the light saturation region. If the rate of photosynthesis is to be raised in this region, factors other than light intensity would have to be adjusted. The amount of sunlight falling on a clear summer day in many places on the earth is about 100 000 lux, also equal to about 1000 Wm^{-2}. Thus, except for plants growing in thick forests and in shade, there is often sufficient sunlight incident on the plants to saturate their photosynthetic capacity. The energy of the extreme blue (400 nm) and red (800 nm) light quanta differs only by a factor of two and the photons in this wavelength range are qualitatively efficient to start photosynthesis though, as we shall later see, the leaf pigments preferentially absorb light of certain definite wavelengths.

2.3.2   Effect of temperature

A comparison of the curves A and B in the figure shows that at low light intensities the rate of photosynthesis is the same at 15 °C and at 25 °C. The reactions in the light-limiting region, like true photochemical reactions, are not sensitive to temperature. At higher light intensities, however, the rate of photosynthesis is much higher at 25 °C than at 15 °C. So, factors other than mere photon absorption influence photosynthesis in the light-saturation region. Most temperate climate plants function well between 10 °C and 35 °C, the optimum temperature being around 25 °C.

2.3.3   Effect of CO_2 concentration: CO_2 compensation point

In the light-limiting region the rate of photosynthesis is not affected by lowering the CO_2 concentration, as shown by curve C in Fig. 2–3. Thus, it can be inferred that CO_2 does not participate directly in the photochemical reaction. But at light intensities above the light-limiting region, photo-synthesis is appreciably enhanced by increasing the CO_2 concentration. Photosynthesis by some crop plants, in short-term experiments, increased linearly with increasing CO_2 concentration up to about 0.5% though con-tinued exposure to this high CO_2 concentration injured the leaves. Very good rates of photosynthesis can be obtained with a CO_2 content of about 0.1%. The average CO_2 content of the atmosphere is about 0.03 to 0.04%. Therefore plants in their normal environment do not have enough CO_2 to make maximum use of sunlight falling on them. When a leaf kept in an enclosed space is illuminated with saturating light the concentration of CO_2 in the air falls gradually and reaches a constant level known as the CO_2 compensation point. At this point the CO_2 uptake by photosynthesis is balanced by the CO_2 released by respiration (dark and light). The CO_2 compensation point varies with species; it is very low (less than 10 ppm) for the C_4 plants (Chapter 6) whereas for the C_3 photosynthetic species the value is greater than 50 ppm (0.005%).

2.4   Light and dark reactions; flashing light experiments

As early as 1905 the British plant physiologist F. F. Blackman inter-preted the shape of the light saturation curves by suggesting that photo-synthesis is a two-step mechanism involving a photochemical or light re-action and a non-photochemical or dark reaction. The dark reaction which is enzymic is slower than the light reaction and hence at high light intensi-ties the rate of photosynthesis is entirely dependent upon the rate of the dark reaction. The light reaction has a low or zero temperature coefficient while the dark reaction has a high temperature coefficient, characteristic of enzymic reactions. It should be clearly understood that the so-called dark reaction can proceed both in light and in darkness.

The light and dark reactions can be separated by using flash illuminations lasting fractions of a second. Light flashes lasting less than a millisecond (10^{-3} s) can be produced either mechanically by placing a slit in a rotating disc in the path of a steady light beam, or electrically by loading up a con-denser and discharging it through a vacuum tube. Ruby lasers emitting red light at 694 nm are also used as a radiation source. In 1932 Emerson and Arnold illuminated suspensions of *Chlorella* cells with condenser flashes lasting about 10^{-5} s. They measured the rate of oxygen evolution in relation to the energy of the flashes, the duration of the dark intervals be-tween the flashes, and the temperature of the cell suspension. Flash saturation occurred in normal cells when one molecule of O_2 evolved from 2500 chlorophyll molecules. Emerson and Arnold concluded that the maximum yield of photosynthesis is not determined by the number of chlorophyll molecules capturing the light but by the number of enzyme molecules which carry out the dark reaction. They also observed that for dark time intervals (between successive flashes) greater than 0.06 s the yield of oxygen per flash was independent of the dark time interval; the yield per light flash increased with dark time intervals from 0 to 0.06 s. Thus the dark reaction determining the saturation rate of photosynthesis takes about 0.06 s for completion. The average dark reaction time was calculated to be about 0.02 s at 25°C.

2.5 Important discoveries and formulations

The state of knowledge in the field of photosynthesis at the turn of this century could have been represented by the equation

$$CO_2 + H_2O \xrightarrow[\text{chlorophyll}]{\text{light}} (CH_2O) + O_2 \quad (\varDelta G = 48 \times 10^4 \text{ J (114 kcal)})$$

Prior to about 1930 many investigators in the field believed that the primary reaction in photosynthesis was splitting of carbon dioxide by light to carbon and oxygen; the carbon was subsequently reduced to carbohydrates by water in a different series of reactions. Two important discoveries in the 1930s changed this viewpoint. Firstly, a variety of bacterial cells were found to assimilate CO_2 and synthesize carbohydrates without the use of light energy. Then the Dutch microbiologist van Niel in comparative studies of plant and bacterial photosynthesis showed that some bacteria can assimilate CO_2 in light without evolving O_2. Such bacteria would not grow photosynthetically unless they were supplied with a suitable hydrogen donor substrate. Photosynthesis could be represented, according to van Niel by the general equation

$$CO_2 + 2H_2A \xrightarrow[\text{chlorophyll}]{\text{light}} (CH_2O) + H_2O + 2A$$

where H_2A is the oxidizable substrate. Van Niel suggested that photosynthesis of green plants and algae is a special case in which H_2A is H_2O and 2A is O_2. The primary photochemical act in plant photosynthesis would be the splitting of water to yield an oxidant (OH) and a reductant (H). The primary reductant (H) could then bring about the reduction of CO_2 to cell materials and the primary oxidant (OH) could be eliminated through a reaction to liberate O_2 and reform H_2O. The overall equation of photosynthesis for green plants, after van Niel, is

$$CO_2 + 4H_2O \xrightarrow[\text{chlorophyll}]{\text{light}} (CH_2O) + 3H_2O + O_2$$

which is a sum of three individual steps:

(*i*) $4H_2O \xrightarrow[\text{green pigments}]{\text{light}} 4(OH) + 4H$

(*ii*) $4H + CO_2 \longrightarrow (CH_2O) + H_2O$

(*iii*) $4(OH) \longrightarrow 2H_2O + O_2$

The reaction sequences clearly show that the oxygen is evolved from water and not from CO_2.

The second important observation was made in 1937 by R. Hill of Cambridge University. Hill separated the photosynthesizing particles (chloroplasts) of green leaves from the respiratory particles by differential centrifugation of a homogenate of leaf tissues. Hill's chloroplasts did not evolve O_2 when illuminated as such (due to possible damage of the chloroplasts during isolation) but did so when suitable electron acceptors (oxidants) like potassium ferrioxalate or potassium ferricyanide were added to the illuminated suspension. One molecule of O_2 was evolved for every four equivalents of oxidant reduced photochemically. Later many quinones and dyes were found to be reduced by illuminated chloroplasts. The chloroplasts, however, failed to reduce CO_2, the natural electron acceptor of photosynthesis. This phenomenon, now known as the Hill reaction, is a light-driven transfer of electrons from water to non-physiological oxidants (Hill reagents) against the chemical potential gradient. The significance of the Hill reaction lies in the demonstration of the fact that photochemical O_2 evolution can be separated from CO_2 reduction in photosynthesis.

The decomposition of water, and the resulting liberation of O_2 during photosynthesis, was established by Ruben and Kamen in California in 1941. They exposed photosynthesizing cells to water enriched in oxygen isotope of mass 18 (^{18}O). The isotopic composition of the oxygen evolved was the same as that of the water and not that of CO_2 used. Kamen and Ruben also discovered the radioactive isotope ^{14}C which was successfully used by Bassham, Benson and Calvin in California to trace the path of carbon in photosynthesis (Chapter 6). Calvin and co-workers showed that reduction of CO_2 to sugars proceeded by dark enzymic reactions and also that two molecules of reduced pyridine nucleotide [$NADPH_2$] and three molecules of ATP were required for the reduction of every molecule of CO_2. The role of pyridine nucleotides and ATP in the respiration of tissues was already established by this time. The photosynthetic reduction of NADP to $NADPH_2$ by isolated chloroplasts with simultaneous evolution of O_2 was demonstrated in 1951 by three different laboratories. Arnon, Allen and Whatley in 1954 also demonstrated cell-free photosynthesis, i.e. the assimilation of CO_2 and evolution of O_2 by isolated spinach chloroplasts. Proteins like ferredoxin, plastocyanin, ferredoxin-NADP reductase and cytochromes *b* and *f*, which participate in the transfer of electrons in photosynthesis, were isolated from chloroplasts within a decade.

To conclude, healthy green leaves on illumination generate $NADPH_2$ and ATP. The reducing power of $NADPH_2$ and the energy of hydrolysis of ATP are utilized to reduce CO_2 to carbohydrates in the presence of enzymes some of whose activities are regulated by light.

3 Photosynthetic Apparatus

The photosynthetic apparatus is that part of the leaf or algal cell which contains the ingredients for absorbing light and for channelling the energy of the excited pigment molecules into a series of chemical and enzymatic reactions. Engelmann's experiments (Chapter 2) have shown that chlorophylls are the pigments responsible for capturing light quanta. Knowledge concerning the sub-cellular structure in which chlorophyll is located comes from light and electron microscopy, and from cell fractionation techniques. In green algae and in higher plants the chlorophyll is contained in a cellular plastid called the *chloroplast*. Electron microscope pictures show that chloroplasts in higher plants, e.g. spinach, tobacco, are saucer-shaped bodies 4 to 10 μm in diameter and 1 μm in thickness (1 μm = 10^{-6}m) with an outer membrane or envelope separating it from the rest of the cytoplasm (Fig. 3–1). The number of chloroplasts per cell, in higher plants, varies from one to more than a hundred depending upon the particular plant and on the growth conditions. In many plants the chloroplasts are able to reproduce themselves by a simple division.

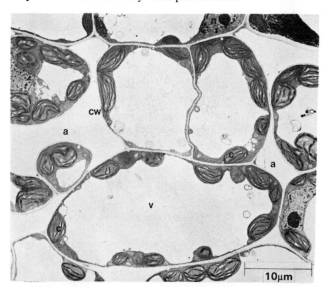

Fig. 3–1 Thin section of spinach mesophyll cells showing chloroplasts (c) in cytoplasm extending around the inside of the cell wall (cw). n, nucleus; v, vacuole; a, air space between cells allowing easy diffusion of gases to chloroplasts. (Courtesy A. D. Greenwood, Department of Botany, Imperial College, London.)

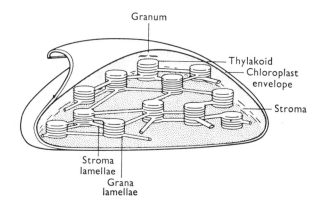

Granum

Thylakoid
Chloroplast
envelope

Stroma

Stroma
lamellae

Grana
lamellae

(a)

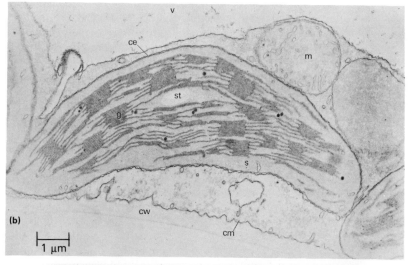

(b)

1 μm

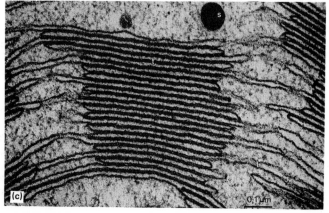

(c)

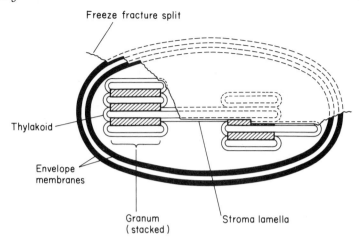

Freeze fracture split

Thylakoid

Envelope
membranes

Granum
(stacked)

Stroma lamella

Fig. 3–3 Diagram of chloroplast structure as revealed by freeze etching in the electron microscope. Freeze fracturing of the chloroplast results in cross-fracturing or splitting of the lamellae, thus exposing interior regions of the membrane (after Staehelin, L. A., 1976, *J. Cell Biol.*, **71**, 136).

Internally the chloroplast is comprised of a system of lamellae or flattened *thylakoids* which are arranged in stacks in certain dense green regions known as *grana* (Fig. 3–2). Each lamella in the chloroplast may contain two double-layer membranes. The grana are embedded in a colourless matrix called the *stroma* and the whole chloroplast is surrounded by a bounding double membrane, the *chloroplast envelope*. Within a chloroplast the grana are interconnected by a system of loosely arranged membranes called the stroma lamellae. The detailed structure of the thylakoids is shown in Figs 3–3 and 3–4. These models are based on electron microscopy using freeze-fracturing techniques, illustrated in Fig. 3–3. The surfaces thus exposed show the distribution of chlorophyll-protein complexes embedded in or associated with the lipid bilayer which forms the 'backbone' of the membrane. The organization of these complexes which are seen as particles is

Fig. 3–2 (a) Cut-away representation of a chloroplast to show three-dimensional structure. (b) Section of a chloroplast in the cytoplasm of a spinach leaf cell. ce, chloroplast envelope; g, granum consisting of stacks of thylakoids; s, stroma; st, starch granule in chloroplast; cm, cytoplasmic membrane; cw, cell wall; m, mitochondrion; v, vacuole. (c) A single granum within a chloroplast showing stacks of thylakoids and interconnecting stroma lamellae between granal stacks. s, lipid droplet in the stroma. (Courtesy A. D. Greenwood.)

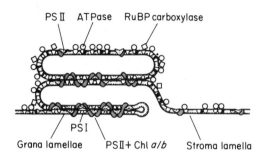

Fig. 3-4 Model of supra molecular organization of thylakoid membranes showing also the different composition in the stroma and grana lamellae (after Staehelin, L. A., and Arntzen, C. J., 1979, *CIBA Fndn. Symp.*, **61**, 147).

different in the stacked, grana membrane regions than in the unstacked, stroma membranes. This is explained by predominance of the photosystem II, oxygen evolving reaction complexes in the grana and the photosystem I particles in the stroma lamellae (see Fig. 5-2).

The lamellar structure is found not only in chloroplasts of higher plants but also in algal chloroplasts. In algae the shapes of the chloroplast are varied Figure 3-5 show chloroplasts from a green and a red alga. The most primitive algae, the blue-greens (also called Cyanobacteria) do not contain chloroplasts as such. The photosynthetic material present in these organisms consists of parallel layers of lamellar membranes traversing the cytoplasm.

It is possible to fractionate the chloroplasts of higher plants so that the green lamellae are separated from the colourless stroma matrix. The lamellar membranes in which the chlorophyll is embedded are approximately half lipid and half protein in chemical composition. The proteins catalyze the enzyme reactions and give mechanical strength to the membranes. Most of the light-harvesting chlorophylls *a* and *b* are conjugated with specific membrane proteins. The presence of lipids facilitates energy storage and offers selective permeability of sugars, salts, substrates, etc. Chloroplast lipids play an important role in maintaining membrane structure and function. One of the causes for the thermal and photo decay of chloroplasts is the release of lipids from the membranes and their oxidation.

The quantum conversion of light energy and associated electron transport reactions of photosynthesis occur in the lamellae. The stroma contains many soluble proteins including the enzymes of the Calvin–Benson cycle

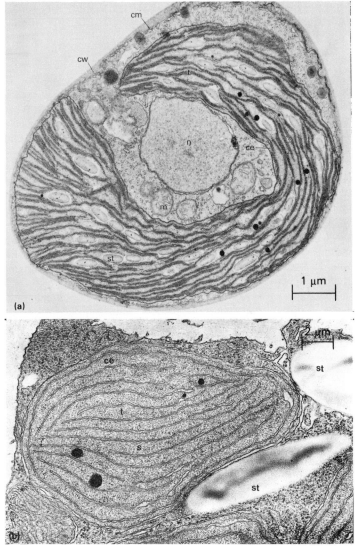

Fig. 3-5 The green alga *Coccomyxa* sp., which is a symbiont within a lichen, showing a single cup-shaped chloroplast within the cell. The thylakoids (t) are in groups of three. ce, chloroplast envelope; st, starch granule within chloroplast; cw, cell wall; cm, cytoplasmic membrane; n, nucleus; m, mitochondrion. (Courtesy H. Bronwen Griffiths, Department of Botany, Imperial College, London.) (b) Chloroplast of the red alga *Ceramium* sp. with single thylakoids (t) lying nearly parallel to one another in the stroma (s). ce, chloroplast envelope; st, starch granule outside chloroplast. Dark spots in chloroplast are lipid droplets. (Courtesy A. D. Greenwood.)

(Chapter 6) which carry out the dark phase reduction of CO_2 to carbohydrates.

3.1 Isolation of chloroplasts from leaves

The first photosynthetically active chloroplasts were isolated by Hill; these preparations were active only in O_2 evolution coupled to the reduction of non-physiological electron acceptors (see Chapter 2). Arnon and Whatley isolated chloroplasts in isotonic sodium chloride, i.e. about 0.35 M or 2%. Their preparations were capable of photoreduction of NADP and photophosphorylation but were able to fix CO_2 only at low rates, although they contained all the enzymes of the Calvin–Benson CO_2 fixation cycle. These chloroplasts appeared intact under the light microscope but electron microscope pictures of the preparation indicated that they had lost their outer membranes and were naked lamellar systems. Such preparations are called 'Type C' (broken) chloroplasts (Fig. 3–6). Walker has developed techniques for the isolation of 'Type A' (complete) chloroplasts which retain their outer envelopes and which can fix CO_2 at rates up to 90% of those of whole leaves. See HALL (1972) *Nature*, **235**, 125 for discussion of chloroplast Types.

Two methods used in our laboratory for the preparation of chloroplasts are given below. All solutions and apparatus used should be pre-cooled in ice. The preparation should be carried out as quickly as possible. See also Chapter 5.

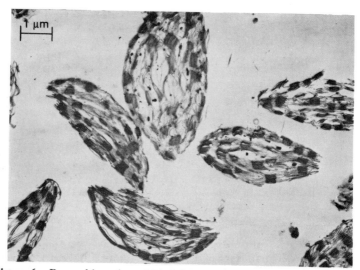

Fig. 3–6 Bean chloroplasts isolated in buffered sucrose media showing chloroplasts with intact thylakoids but without a chloroplast envelope— Type C as defined in chapter 3. (Courtesy of A. D. Greenwood and R. Leech.)

Preparation 1   Procedure of Whatley and Arnon (modified).
   Grinding medium: 0.35 M NaCl; 0.04 M tris/HCl buffer pH 8.0

   Cut 25 g of spinach leaves into small pieces 0.5–1 cm long. Place in an
M.S.E. Atomix (or domestic blender) with 50 cm^3 grinding medium.
Blend for 10 s on low and 20 s on high speed. Filter the homogenate
through a nylon bag (or 4 layers of cheesecloth) into a centrifuge tube.
Centrifuge 4 min at 2000 g. Discard the supernatant. Re-suspend the
pellet in 2 cm^3 0.35 M NaCl using a small piece of absorbent cotton wool
wrapped round the end of a glass rod. This preparation consists of Type C
(broken) chloroplasts. Chloroplast fragments (Type E) are prepared by
diluting the suspension with 10 vol of water to give a NaCl concentration
of 0.035 M.

Preparation 2   Procedure of Walker.
   Grinding medium:

Sorbitol	0.33 M
MgCl$_2$	0.005 M
Na$_4$P$_2$O$_7 \cdot$10H$_2$O	0.01 M

Adjust pH of above mixture to 6.5 with HCl and add sodium
isoascorbate to a final concentration of 0.002 M just before use.

   Re-suspending medium:

Sorbitol	0.33 M
MgCl$_2$	0.001 M
MnCl$_2$	0.001 M
EDTA (ethylenediamine tetra acetate)	0.002 M
HEPES (hydroxyethylpiperazine-ethanesulphonic acid – buffer)	0.05 M.

Adjust pH to 7.6 with NaOH.
   Homogenize 50 g of chilled spinach leaves with 200 cm^3 of freshly-
made grinding medium for 3 to 5 s in a domestic blender. Squeeze
the macerate through 2 layers of cheesecloth and filter through 8 layers
of cheesecloth into 50 cm^3 plastic centrifuge tubes. Centrifuge rapidly
at 0 °C from rest to 4000 g to rest in approximately 90 s. Re-suspend
the pellet gently using a glass rod and a small piece of absorbent cotton
in 1 cm^3 of resuspending medium. This procedure should produce a
suspension of chloroplasts, 50–80%, Type A (complete), which would be
capable of high rates of CO$_2$ fixation.

3.2   Chloroplast pigments

   All photosynthetic organisms contain one or more organic pigments
capable of absorbing visible radiation which will initiate the photochemical
reactions of photosynthesis. These pigments can be extracted from most
leaves into alcohol or into other organic solvents. From the alcoholic
extract individual pigments can be separated by chromatography on a
column of powdered sugar, as was shown by the Russian botanist Tswett

Table 2   The pigments.

Type of pigment	Characteristic absorption maxima (nm) (in organic solvents)	Occurrence
Chlorophylls		
Chlorophyll *a*	420, 660	All higher plants and algae
Chlorophyll *b*	435, 643	All higher plants and green algae
Chlorophyll *c*	445, 625	Diatoms and brown algae
Chlorophyll *d*	450, 690	Red algae
Carotenoids		
β-carotene	425, 450, 480	Higher plants and most algae
α-carotene	420, 440, 470	Most plants and some algae
Luteol	425, 445, 475	Green algae, red algae and higher plants
Violaxanthol	425, 450, 475	Higher plants
Fucoxanthol	425, 450, 475	Diatoms and brown algae
Phycobilins		
Phycoerythrins	490, 546, 576	Red algae and in some blue-green algae (cyanobacteria)
Phycocyanins	618	Blue-green algae and in some red algae
Allophycocyanins	650	Blue-green and red algae

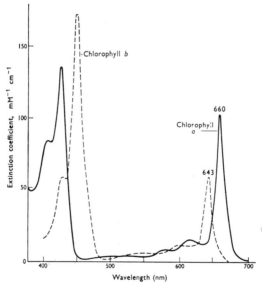

Fig. 3-7   Absorption spectra of chlorophylls extracted in ether. (Redrawn from Zscheile and Comar, 1941, *Bot. Gaz.*, **102**, 463.)

in 1906. The three major classes of pigments found in plants and algae are the chlorophylls, the carotenoids and the phycobilins. The chlorophylls and carotenoids are insoluble in water but the phycobilins are soluble in water. The carotenoids and phycobilins are called the accessory photosynthetic pigments since the quanta absorbed by these pigments can be transferred to chlorophyll. Table 2 gives the absorption characteristics of these pigments. The photosynthetic pigments of bacteria are discussed in Chapter 7.

Chlorophylls are the pigments that give plants their characteristic green colour. They are insoluble in water but soluble in organic solvents. Chlorophyll a is bluish-green and chlorophyll b is yellowish-green. Chlorophyll a is present in all photosynthetic organisms which evolve O_2. Chlorophyll b is present (about one third of the content of chlorophyll a) in leaves of higher plants and in green algae. The absorption maxima of chlorophyll a and chlorophyll b in ether are respectively at 660 and 643 nm as shown in Fig. 3–7; in acetone the peaks are at 663 and 645 nm. However, careful spectroscopic investigations of the living cell indicate the presence of multiple forms of chlorophyll a in vivo. These forms of chlorophyll a may be associated in different ways with the lamellae and have different photochemical functions.

The molecular formula for chlorophyll a is $C_{55}H_{72}N_4O_5Mg$ and for chlorophyll b is $C_{55}H_{70}N_4O_6Mg$. The structural formula of chlorophyll

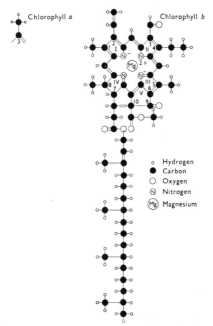

Fig. 3–8 Chlorophyll a and b structure. (Redrawn from HEATH, 1969.)

was determined by Fischer in Germany in 1940 from degradative studies; the structure was confirmed by the complete synthesis of the molecule by Woodward at Harvard in 1960. The chlorophyll molecule (Fig. 3–8) contains a porphyrin 'head' and a phytol 'tail'. The polar (water soluble) porphyrin nucleus is made up of a tetrapyrrole ring and a magnesium atom. In the cell, electron microscopists think that the chlorophyll is sandwiched between protein and lipid layers of the chloroplasts lamellae. The porphyrin part of the molecule is bound to the protein while the phytol chain extends into the lipid layer since it is soluble in lipids.

The optical absorption curves of chlorophyll a and chlorophyll b intersect at 652 nm. A solution of chlorophyll at a concentration of 1 mg cm^{-3} has an optical density of 34.5 at 652 nm. Arnon has devised the following method for the determination of the chlorophyll content of a chloroplast suspension by measuring its absorption at 652 nm. Dilute 0.1 cm^3 of chloroplast suspension with 20 cm^3 of 80% acetone, mix, and filter. Read the absorbancy of the filtrate at 652 nm. in a spectrophotometer in a 1 cm light path cell against 80% acetone as reference. Multiply the absorbance by 5.8 to give mg of chlorophyll per cm^3 of the original chloroplast suspension.

The carotenoids are yellow or orange pigments found in all photosynthesizing cells. Their colour in the leaves is normally masked by chlorophyll, but in the autumn season when chlorophyll disintegrates the yellow pigments become visible. Carotenoids contain a conjugated double bond system of the polyene type. They are usually either hydrocarbons (carotenes) or oxygenated hydrocarbons (carotenols or xanthophylls) of 40 carbon chains built up from isoprene subunits (Fig. 3–9). They have triple-banded absorption spectra in the region from about 400 to 550 nm. The carotenoids are situated in the chloroplast lamellae in close proximity to the chlorophyll. The energy absorbed by the carotenoids may be transferred to chlorophyll a for photosynthesis. In addition the carotenoids may protect the chlorophyll molecules from too much photo-oxidation in excessive light.

Blue-green algae and red marine algae contain a group of pigments known as *phycobilins* (Fig. 3–10). Phycobilins are linear tetrapyrroles structurally related to chlorophyll a but they do not have the phytyl side chain, nor do they contain magnesium. The chromophores of phycobilins are covalently linked to polypeptides to form water-soluble phycobiliproteins. There are three classes of phycobilins viz. *phycoerythrins, phycocyanins* and *allophycocyanins*. The red phycoerythrins found in all red algae (Table 2) absorb light in the middle of the visible spectrum. This enables the red algae living under the sea to perform photosynthesis in the dim bluish-green light reaching the lower surfaces of the ocean – the deeper under the sea a red alga lives the more phycoerythrin it contains in relation to chlorophyll. The blue phycocyanins and allophycocyanins occur in the blue-green algae which live on the surface layers of lakes and on land. The energy absorbed by the phycobilins (accessory pigments) is transferred to the chlorophyll for photochemical processes. Using pico (10^{-12}) second

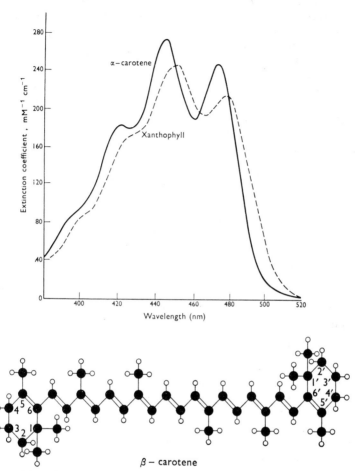

(a)

(b) β – carotene

Fig. 3–9 (a) Absorption spectra of α-carotene and a xanthophyll. (Redrawn from ZSCHEILLE *et al.* (1942), *Plant Physiol.*, **17**, 331.) **(b)** Structure of β-carotene. (Redrawn from HEATH, 1969.)

spectroscopy it has been shown that the energy transfer in the red alga, *Porphyridium cruentum* occurs in the sequence:

phycoerythrin —→ phycocyanin —→ allophycocyanin —→ chlorophyll *a*.

Thus higher plants and algae, during the course of evolution, have developed various pigments to capture the available solar radiation most efficiently and to carry out photosynthesis. The relative abundance of these pigments depends upon the species, the location of the plant, the seasons, etc.

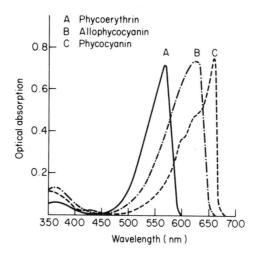

Fig. 3–10 Absorption spectra of phycobiliproteins in the visible region (after Bennet, A. and Bogorad, L., 1971, *Biochemistry*, **10**, 3625).

In addition to the pigments the lamellae are composed of many proteins, lipids, quinones and metal ions. The role of some of these constituents in photosynthesis has been established by the technique of difference absorption spectroscopy (see Chapter 4). Two cytochromes, cytochrome b_6 and cytochrome f, found in chloroplasts are involved in photosynthetic electron transport. The blue copper-containing protein plastocyanin, the non-heme iron protein ferredoxin and the flavoprotein ferredoxin-NADP reductase are also located in the chloroplasts. Plastoquinone is believed to be involved in the earlier stages of electron transfer from excited chlorophyll molecules. Zinc, iron, magnesium and manganese are some of the metal ions found in chloroplast lamellae.

3.3 The photosynthetic unit

Functionally chlorophyll molecules act in groups. A photosynthetic unit is conceived as a group of pigments and other molecules utilizing the transfer of excitation energy as a mechanism by which the reaction centre communicates with an antenna of light-harvesting pigments as shown in Fig. 3–11. According to this concept a single quantum of energy absorbed anywhere in a set of about 250 chlorophyll molecules migrates to a reaction centre containing a special pair of chlorophyll a molecules and promotes an electron transfer event. The following observations lead to the idea of the existence of a photosynthetic unit.

1. Approximately 8 quanta of light absorbed by chlorophyll are required for the photosynthetic reduction of 1 CO_2 molecule and the evolu-

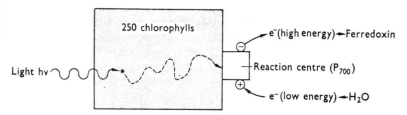

Fig. 3–11 Diagrammatic representation of the photosynthetic unit. One photon of light reacts in a unit of 250 chlorophyll molecules containing one P_{700} reaction centre.

tion of 1 O_2 molecule. If each chlorophyll molecule in a plant can react photochemically a sufficiently intense flash of light should bring about the evolution of 1 O_2 for every 8 chlorophyll molecules present. However, the flashing light experiments of Emerson and Arnold (Chapter 2) on *Chlorella* suspensions showed that the maximum yield per flash was 1 O_2 molecule for about 2500 chlorophyll molecules, that is a quantum of light is absorbed by one in a cluster of about 300 chlorophyll molecules.

2. Gaffron and Wohl calculated that a single chlorophyll molecule in a dimly illuminated plant will absorb a light quantum only once in several minutes. At this rate a single molecule of chlorophyll will require nearly an hour to capture the light quanta needed for the evolution of one molecule of O_2. But when a plant is illuminated the maximum rates of CO_2 uptake and O_2 evolution are quickly established. So Gaffron and Wohl postulated that the energy harvested by a large set of chlorophyll molecules is conducted to a single reaction centre.

3. Gaffron and co-workers also observed that certain golden-yellow leaves of tobacco, with very little chlorophyll, may reach a nearly normal rate of photosynthesis at very high light intensities. These leaves would have contained a larger proportion of the special type of chlorophyll molecule which is in direct touch with the components of the electron transfer chain.

4. Difference spectroscopy has revealed the role of certain special components like P_{700} (by Kok in 1956) and cytochrome (by Duysens in 1961), in the photochemical electron transfer reactions. There is one molecule of light-reacting cytochrome and one P_{700} for every 250 chlorophyll molecules in higher plants and algae.

3·4 Photosynthetic apparatus of C_4 plants

Leaves of plants like sugar cane, maize (corn), *Sorghum*, *Amaranthus* and many tropical grasses contain two distinct types of chloroplasts. The leaves of these plants possess 'Kranz-type' anatomy (kranz in German means wreath). The chloroplasts are located in the leaf cells in two concentric layers surrounding the vascular bundle; the inner layer is called the *bundle-sheath* and the outer the *mesophyll*. These chloroplasts also possess a membrane system in the peripheral stroma, the peripheral reticulum, which connects the thylakoid membranes to the chloroplast envelope. The two types of chloroplasts can be separated by careful grinding and centrifugation using density gradients. As will be discussed in Chapter 6 these plants are able to fix CO$_2$ by two different pathways: (a) the normal Calvin cycle in the bundle-sheath chloroplasts where the initial product of CO$_2$ fixation is the three carbon compound phosphoglyceric acid (C$_3$ pathway) and (b)

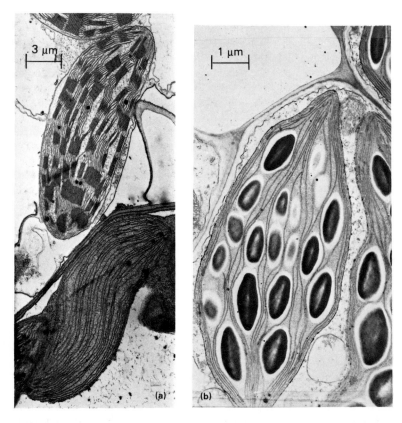

Fig. 3–12 (a) Two different types of chloroplasts in the maize leaf (a C$_4$ plant). Above, mesophyll or granal type. Below, bundle sheath or agranal type. There are no starch grains in the bundle sheath chloroplast as the leaf was kept in darkness for 24 hours before fixation. (Courtesy G. Montes, King's College, London.) (b) Bundle sheath (agranal) chloroplast from maize. (From WHATLEY and WHATLEY, 1980.)

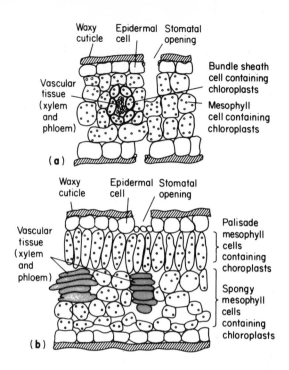

Fig. 3-13 Differences in chloroplast structure in C_4 (**a**) and C_3 (**b**) species. Note the absence of bundle sheath cells in the C_3 species. (After Zelitch, I., Feb. 5, 1979, *Chemical and Engineering News, U.S.A.*, 37.)

in the mesophyll chloroplasts by combining CO_2 with phosphoenol pyruvate to produce the four carbon acids oxaloacetate and malate (C_4 pathway). Plants which fix CO_2 only by the Calvin cycle are termed C_3 species and are generally grown in the temperate zones, e.g. wheat, spinach, oak tree, and those plants which fix CO_2 both by the Calvin cycle and by the malate pathway are termed C_4 species and are generally adapted to growth in warmer and/or drier climates as found in hotter zones.

The mesophyll chloroplasts of C_4 plants (Fig. 3-12a) are randomly distributed in the cell, have stacks of grana and few starch granules. The bundle sheath chloroplasts are relatively larger in size, generally lack grana and possess a number of starch granules (Fig. 3-12b). The mesophyll and bundle sheath chloroplasts lie adjacent in the leaf and photosynthetic products can easily flow from one type to the other (see Fig. 3-13a). Generally plants possessing these dimorphic type chloroplasts have a very low CO_2 compensation point, very low photorespiration and glycollate metabolism and grow more rapidly with higher crop yield.

4 Light Absorption and Emission by Atoms and Molecules

The most stable states of atoms are those in which the valence electrons are distributed, in accordance with the Pauli principle, into the quantum states of least energy, i.e. the electrons are in their ground states of energy level. When light is absorbed by an atom in the ground state the whole energy of the quantum ($h\nu$) is added to it, and the electrons are lifted to an energy-rich excited state. This is illustrated in Fig. 4–1, taking the helium atom as an

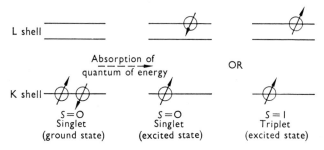

Fig. 4–1 Energy levels of electrons in the helium atom.

example. The time needed for the entire process is of the order of 10^{-15} s. If an atom has an even number of electrons, the spins of these electrons are usually arranged in opposite directions and cancel each other out so that the total electronic spin of the atom s=0 (singlet state). For an atom with an odd number of electrons the net spin is s=$\frac{1}{2}$ (doublet state). If an atom has an even number of electrons and if the electronic spins are in the parallel direction the net spin s=1 and the atom is said to be in the triplet state. These states are illustrated in Fig. 4–1. The transition from the ground to the excited state of an atom by the absorption of a single quantum of energy can be followed by a sharp line in its absorption spectrum at the wavelength λ given by $\Delta E = hc/\lambda$ (see Chapter 1). In a molecule consisting of various atoms the transition from the ground to the excited state can take place by the absorption of light of varying amounts of energy quanta; the sharp line of the atomic absorption spectrum then is replaced by a broad absorption band. In individual atoms absorption and emission take place at the same wavelength, while in a whole molecule the absorption and emission spectrum do not coincide; the peak of the emission spectrum is at a longer wavelength than the peak of the corresponding absorption spectrum (Fig. 4–2a).

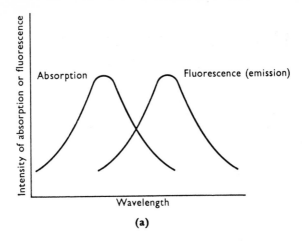

(a)

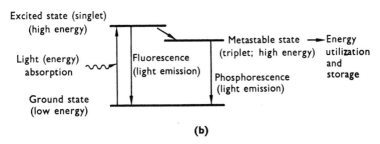

(b)

Fig. 4-2 Spectra of light absorption and emission.

4.1 Time spans involved; fluorescence and phosphorescence

A molecule, say of chlorophyll, in an electronically excited state can revert to the ground state in a number of ways. It can dissipate some of the
acquired energy as heat and emit a photon back into space (Fig. 4–2b). This
phenomenon is called *fluorescence*. The wavelengths of a fluorescence spectrum are always longer than the wavelengths of the corresponding absorption spectrum. Chlorophyll *a* extracts, for example, absorb in blue and red
regions of the spectrum but fluoresce only in the red; the maximum intensity of fluorescence occurs at 668 nm compared to the wavelength of
maximum absorption at 663 nm. The average time a molecule has to spend
in the excited state between absorption and emission is known as the
natural lifetime of the excited state; its duration depends on the electronic
properties of the excited state. The average fluorescence lifetime is of the
order of 10^{-9} s.

A second route by which an excited molecule can lose its energy is by
transfer from its original excited singlet state into a metastable triplet state
with a much longer lifetime (of the order of milliseconds). From the meta-

stable triplet state the molecule can revert to the natural ground state by emitting a photon at a longer wavelength. This weak emission is known as *phosphorescence*. Phosphorescence is slow enough to be observed by the eye even when the exciting light is turned off. Due to their longer lifetime, lower energy, and magnetic moment (since the excited electron and its partner have parallel spins) the *triplet excited states* are of importance in photochemistry. The probability of an excited chlorophyll molecule reacting with another molecule is very much higher in the triplet (phosphorescent) state than in the singlet (fluorescent) state. The existence of triplet states has been shown in chlorophyll dissolved in organic solvents but not yet clearly demonstrated during the normal process of photosynthesis.

4.2   Energy transfer or sensitized fluorescence

The phenomenon of sensitized fluorescence involves the interaction of two molecules that may be separated in solution by many molecules of the solvent. In this type of energy transfer, two kinds of pigments are dissolved in the same solvent and the solution is illuminated with light of such wavelength that can be absorbed by only one of the pigments called the *donor*. The wavelength of light emitted from the solution, however, corresponds to the fluorescence spectrum of the second pigment molecule, the *acceptor*. The energy of excitation of the donor molecule is transferred by resonance to the acceptor molecule. One of the requisites for this type of energy transfer is that the fluorescent state of the donor molecule must have an energy

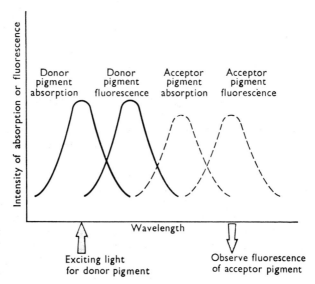

Fig. 4-3   Relationship between absorption and fluorescence spectra of donor and acceptor pigments.

greater than or equal to the fluorescent state of the acceptor molecule, that is, the fluorescent band of the donor molecule should overlap with the absorption band of the acceptor (Fig. 4–3). The quanta taken up by accessory pigments in many algae are transferred either wholly or partly to chlorophyll *a* by sensitized fluorescence. In the green algae chlorophyll *a* fluorescence is observed when light is absorbed by chlorophylls *a* and *b* and also by the carotenoids. Chlorophyll *a* has its fluorescence at the longest wavelength so that the migration of energy is always from the other excited chloroplast pigments to chlorophyll *a*.

From studies of chlorophyll *a* fluorescence and quantum yields of photosynthesis in *Chlorella* it was shown that in this alga the transfer of excitation energy from chlorophyll *b* to *a* is 100% efficient while the energy transfer from the carotenoids to chlorophyll *a* is only 40% efficient. The close assembly of various pigment molecules in the lamellae is necessary for efficient energy transfer.

4.3   Emerson effect and the two-light reactions

A plot showing the efficiency of photosynthesis (measured as O_2 evolution or CO_2 fixation) by monochromatic light as a function of the wavelength of light is known as the *action spectrum* of photosynthesis. For photochemical reactions involving a single pigment the action spectrum has the same general shape as the absorption spectrum of the pigment. If P molecules of O_2 per second are evolved from a system which absorbs I quanta of monochromatic radiation per second, then the ratio P/I, (Φ), is called the quantum yield or *quantum efficiency* of photosynthesis. The reciprocal of the quantum yield $[1/\Phi]$ which gives the number of quanta required to liberate one molecule of O_2 is usually called the *quantum requirement* of photosynthesis. Although values ranging from 4 to 12 have been mentioned for the quantum requirement by various workers in the field the more widely accepted value is a minimum of 8.

Emerson and associates at the University of Illinois in the 1940s studied the action spectra of photosynthesis for various algae by measuring the maximum quantum yield of photosynthesis as a function of the monochromatic light used to illuminate the algae. They found that the most effective light for photosynthesis, in *Chlorella*, was red (650 to 680 nm) and blue (400 to 460 nm), those colours that are most strongly absorbed by chlorophyll. The photosynthetic efficiency of a quantum absorbed at 680 nm was about 36% more than that of a quantum absorbed at 490 nm.

The quantum yield of photosynthesis decreased very dramatically with increasing wavelength beyond 685 nm even though chlorophylls still absorb light at these wavelengths. This fact, the so-called *red drop* in photosynthesis, could not be explained at that time. However, Emerson and co-workers later showed that the amount of photosynthesis in far red

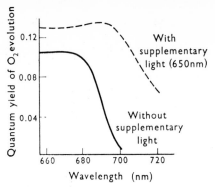

Fig. 4–4 Efficiency of photosynthesis (quantum yield) in the green alga *Chlorella* at different wavelengths of light. On adding supplementary light the quantum yield is enhanced at wavelengths above 680 nm – the Emerson enhancement effect. (Redrawn from Emerson *et al.* (1957) *Proc. Natl. Acad. Sci. U.S.*, **43**, 133.)

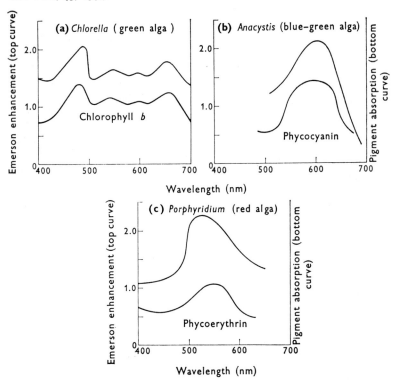

Fig. 4–5 Action spectrum of the Emerson effect in different algae (upper curve in each case) correlated with the absorption of the accessary pigments of the same algae (lower curve). (a) *Chlorella* containing chlorophyll *b*; (b) *Anacystis* containing phycocyanin; (c) *Porphyridium* containing phyco-erythrin. (Redrawn from Emerson and Rabinowitch, 1960, *Plant Physiol.*, **35**, 477.)

light (wavelengths greater than 685 nm) could be increased considerably by a supplementary beam of red light (about 650 nm) (see Fig. 4–4). In fact the total amount of photosynthesis carried out in the presence of a mixture of far red light and red light was greater than the sum of the amounts of photosynthesis carried out in separate experiments with the individual beams of light. This increase of the photosynthetic efficiency of far red light in the presence of a supplementary beam of lower wavelength is known as the *Emerson enhancement effect.* Experimental results of this effect are shown for three plants with different accessory pigments in Fig. 4–5. What is actually measured is O_2 evolution. It will be noticed that the simultaneous illumination of the plant by light of two different wavelengths results in greater O_2 evolution than the sum of that yielded by the two wavelengths applied separately. In other words, enhancement has occurred. Thus plant photosynthesis involves two light reactions; one sensitized by chlorophyll *a* and one sensitized by accessory pigments – these are now termed Photosystem I and Photosystem II respectively.

Important subsequent work by Myers and French in 1960 showed that Photosystem I and Photosystem II need not be applied simultaneously to get optimum photosynthesis but that they may be given alternately with a short dark period of a few seconds between them. This indicated that the two-light reactions could store their photochemical products for a short time before reacting with the electron transfer chain.

Experimental evidence and theoretical postulates in support of the two-light reaction hypothesis followed. The difference in electrode potential (ΔE) between the reactants and the final products in the photosynthetic reaction is 1.25 V (ΔE of CO_2 – glucose couple = -0.43 V; ΔE of H_2O – O_2 couple = $+0.82$ V). But since 4 electrons are required to evolve one

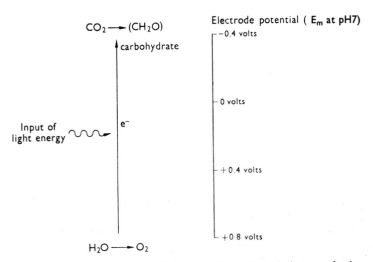

Fig. 4–6   Redox potentials of the overall reaction of photosynthesis.

molecule of O_2 and reduce one molecule of CO_2 the total energy requirement is $1.25 \times 4 \times 9.64 = 48.2 \times 10^4$J (see pp. 6, 14 and 60). Interestingly,
this energy gap is greater than that which can be overcome by the absorption of two quanta of photosynthetically-competent red light which has an
energy content of 17.10×10^4J mole^{-1}. If however, a quantum requirement
of 8 is accepted as the value for photosynthesis (p. 33), then two light
reactions would be needed for each O_2 evolved since one quantum of light
can only activate one electron at a time (therefore 8 quanta are required per
O_2 per 4 electrons, i.e. 2 quanta per electron). In 1960 Hill and Bendall
at Cambridge put forward the idea that the two-light reactions should
be in series (and not in parallel) with cytochromes b_6 and f acting as
electron carriers in the 'dark' reaction which connects the two photosystems. This is shown in Fig. 4–7 and is elaborated and updated in
Chapter 5. The most conclusive evidence for two separate photosystems

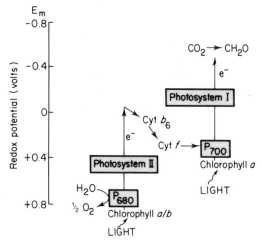

Fig. 4–7   The two-light reaction or 'Z' scheme of photosynthesis. (Concept
of Hill and Bendall, 1960, *Nature*, **186**, 136.)

came from a series of difference spectroscopy studies initiated by Duysens
and by Kok and elaborated further by Witt. In this type of measurement
the change in absorbancies of various constituents of a cell suspension are
studied by illumination with mono-chromatic light of varying wavelenths.
A schematic diagram of a difference spectrophotometer is shown in Fig.
4–8. By following the qualitative and quantitative changes in the difference
spectrum of individual photosynthetic reaction components, e.g. cytochromes, the role of some of these components in the electron transfer
pathway can be inferred. For example, Duysens illuminated a suspension
of the red alga *Porphyridium* in the presence of DCMU (a synthetic weed
killer which inhibits oxygen evolution) and found an accumulation of

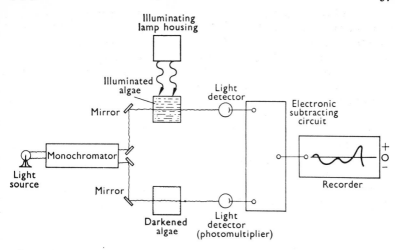

Fig. 4–8 Diagram of a difference spectrophotometer to measure the absorption of pigments.

cytochrome f in the oxidized form. When the experiment was repeated with unpoisoned alga there was no change in the cytochrome f spectrum. Duysens concluded that cytochrome f was an intermediate in the reactions accompanying oxygen evolution (see also Chapter 5).

4.4   Reaction centres and primary electron acceptors

By illuminating algae suspensions with brief flashes of light Kok was able to identify and characterize a special type of chlorophyll a which he called P_{700}. P_{700} is a trace constituent of chloroplasts with an absorption peak around 700 nm. It is reversibly bleached in light; the bleaching corresponds to an oxidation of P_{700} to P_{700}^+. Subsequent investigations have shown that P_{700} is the reaction centre 'trap' pigment in which all incident light energy of wavelength greater than about 680 nm is captured and utilized for primary photochemical reactions, i.e. P_{700} is the primary electron donor of PSI. P_{700} is believed to be a dimer of specialized chlorophyll a molecules; normally chloroplasts contain one P_{700} molecule for every 200–300 bulk chlorophyll molecules. Light displaces an electron from P_{700} to an acceptor molecule and simultaneously an electron moves from plastocyanin (or from cytochrome f in some algae) to P_{700}^+. The chemical nature of the acceptor of electrons from P_{700}, i.e. the primary electron acceptor of PSI is still not definitely established. It is designated as X. Electron paramagnetic resonance (EPR) investigations of chloroplasts at very low temperatures have shown the existence of two membrane-bound iron-sulphur centres (A and B) with very low midpoint redox potentials (see Chapter 5) immediately following X and preceding ferredoxin in the electron transport pathway.

Photosystem II has a higher content of chlorophyll b than PSI and is strongly fluorescent. The reaction centre chlorophyll of PSII is designated as P_{680}. Witt and co-workers followed chlorophyll absorption changes and rates of O_2 evolution in flashing light experiments in normal and DCMU-treated chloroplasts. They found a correlation between the O_2 evolution rate and the photo-induced absorption change of a chloroplast component with a peak around 682 nm. Further investigations have shown that P_{680} is also a specialized type of chlorophyll a. The primary electron acceptor of PSII is designated as Q (as it is a quencher of the chlorophyll fluorescence). It may be a complex of phaeophytin and quinones. From Q the electrons migrate to a plastoquinone pool which acts as an electron reservoir between the two photosystems (see Chapter 5).

4.5 Photosynthetic oxygen evolution

The water-splitting reaction is carried out on the internal surface of the thylakoid membrane on the oxidizing side of PSII (Fig. 5–2). Photo-excitation of P_{680} generates a strong oxidant P_{680}^+ which promotes the oxidation of the primary donor to PSII (usually designated Z) and the turnover of the water splitting system S. S is considered to be a manganese-protein complex which can accumulate four positive charges – it can cycle through five different oxidation states (S_0, S_1, S_2, S_3, S_4) each differing from the preceding state by the loss of a single electron. Each photoact thus transfers an electron from the reaction centre to the acceptor Q and then an electron is transferred from S to the reaction centre. The water-splitting system S which has lost four electrons is then able to react with water, producing oxygen and returning S to its most reduced state S_0. Though various models for photooxidation of water have been proposed and Mn-proteins reportedly involved in the catalytic centre have been isolated, the mechanism of water-splitting is still not clearly understood.

4.6 Experimental separation of the two photosystems

In recent years various attempts have been made to physically separate PSI and PSII from chloroplasts. Chloroplasts can be fragmented by the addition of detergents like digitonin or sodium dodecyl sulphate or Triton X-100, by sonic vibration or by mechanical extrusion under high pressure (about 800 atmospheres) through a French Press. The fragments are then separated by a combination of differential and density gradient centrifugation and chromatography techniques. Some of the preparations thus obtained had a very high chlorophyll a to chlorophyll b ratio and a high P_{700} content suggesting they are highly enriched in PSI constituents. Particles enriched in PSI or PSII constituents have also been isolated from blue-green algae. However, a complete separation of the active pigment systems has not yet been achieved. It seems to be very difficult to isolate a PSII particle free of PSI particles. Recently plant and algal mutants lacking either PSI or PSII have been produced by genetic manipulation and this should help in the isolation of pure PS particles.

By treatment with detergents the proteins tightly bound to the thylakoid membranes can be brought into solution. The soluble components can then be resolved into fractions of varying molecular weights by electrophoresis on gels, e.g. polyacrylamide (PAGE), and located as specific bands in the gel by staining with dyes. Also it is possible to select mutants of plants and algae lacking specific chloroplast pigments or proteins and often deficient in some key photosynthetic activity or spectral characteristic. By comparing the gel bands from chloroplasts of such mutant and wild type plants it is possible to assign a definite function to an individual protein or pigment complex identified in the gel. Some of the thylakoid components separated and tentatively identified by the above techniques are: a P_{700}-chlorophyll a–protein complex associated with the PSI reaction centre, a light harvesting chlorophyll a/b protein complex associated with the PSII reaction, membrane-bound iron-sulphur proteins, cytochrome f, ATPase-coupling factors, etc. (see Fig. 8–1).

5 Photosynthetic Electron Transport and Phosphorylation

In Chapter 2 we elucidated the idea of photosynthesis involving both light and dark phases in the fixation of CO_2. The recognition and experimental demonstration of these two phases was an important step towards the modern understanding of the CO_2 fixation process. This was not possible until Arnon, Allen and Whatley in 1954 were able to isolate chloroplasts from spinach leaves which were capable of carrying out complete photosynthesis, i.e. fixing CO_2 to the level of carbohydrate (see Chapter 3 for the preparation technique of chloroplasts). They were able to physically separate the light and dark phases and to show the light-dependent formation of ATP and $NADPH_2$, which then acted as the energy sources for the subsequent dark fixation of CO_2. This is summarized in the familiar diagram below (Fig. 5-1).

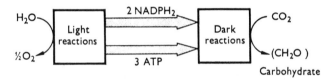

Fig. 5-1 Major products of the light and dark reactions of photosynthesis.

The light phase, which occurs subsequent to the initial light reactions discussed in the previous chapter, involves biochemical reactions with life times of 10^{-1} to 10^{-5} s. The initial light reactions have of course much shorter lifetimes – down to 10^{-9} s. The biochemical events of the light phase results in (*i*) the production of the strong reducing agent, $NADPH_2$, (*ii*) the accompanying evolution of O_2 as a by-product of the splitting of H_2O and (*iii*) the formation of ATP which is coupled to the flow of electrons from H_2O to NADP.

In this chapter we shall discuss how these reactions are thought to occur, what evidence there is for making such assumptions and what compounds are involved in the sequence of electron transport reactions.

5.1 Reduction and oxidation of electron carriers

The production of $NADPH_2$, ATP and O_2 in the lamellae of the chloroplast involves the transfer of electrons through a chain of electron carriers. This electron transfer requires that each of the carriers in turn becomes reduced and oxidized in order that the energy in the electron can be passed along the chain. Reduction simply means the adding of an electron while oxidation implies the removal of an electron from a compound. Whenever an electron is exchanged between two compounds one is oxidized and the other reduced. Almost every such exchange is accompanied by the release or absorption of energy. It makes no difference whether we think of the energy as arising out of the pull exerted on the electron by 'oxidizing power' or the push exerted by 'reducing power'.

Often, though not invariably, an electron travels in company with a proton, i.e. as part of a hydrogen atom. In that case oxidation means removing hydrogen and reduction means adding hydrogen. Thus NADP is reduced to $NADPH_2$, and CO_2 to carbohydrate, by the addition of hydrogen atoms.

The oxidation-reduction ('redox') potentials of biological electron carriers are expressed on a voltage scale at biological pH's which indicates that the $H_2O \rightarrow O_2$ couple is very oxidizing, with a positive midpoint potential of $+0.82$ V while the $H^+ \rightarrow H_2$ (gas) couple is very reducing with a negative potential of -0.42 V. Most biological electron transfer reactions occur between these two extremes (see Table 3). We shall see later that, in fact, the process of photosynthetic electron transport takes place at between $+0.8$ V and -0.4 V. In order to reach these extremes of redox potential light energy is required.

Table 3 Midpoint redox potentials [E_m] of some chloroplast components and reactions.

Chloroplast component or reaction	E_m volts
P_{680} of PSII	$+0.9$ (or more $+$ve)
H_2O/O_2	$+0.82$
P_{700} of PSI	$+0.38$
Cytochrome f	$+0.37$
Platocyanin	$+0.37$
Fe-S centre R	$+0.29$
Platoquinone	0.0
Cytochrome b_6	-0.07
$NADPH_2$	-0.34
H^+/H_2	-0.42
Ferredoxin	-0.43
CO_2/CH_2O	-0.43
Fe-S centre A of PSI	-0.55
Fe-S centre B of PSI	-0.59
Primary e^- acceptor (X) of PSI	-0.7 (or more $-$ve)

5.2 Two types of photosynthetic phosphorylation

Photosynthetic phosphorylation is the production of ATP in the chloroplast by light-activated reactions; it can take place via two systems, non-cyclic and cyclic. In non-cyclic photophosphorylation ATP is generated in an 'open' electron transfer system together with the evolution of O_2 from H_2O and the formation of $NADPH_2$ from NADP. In cyclic photophosphorylation the electrons cycle in a 'closed' system through the phosphorylation sites and ATP is the only product formed. These two systems are shown in Fig. 5–2(a).

5.3 Non-cyclic electron transport

This is the light-requiring process in which low energy electrons are removed from H_2O, resulting in the evolution of O_2 as a by-product (Hill reaction, Chapter 2), and the transfer of these électrons via a number of carriers to produce a strong, negative reducing potential with the subsequent formation of $NADPH_2$, a reducing agent with a potential of -0.34 V. This is simply expressed as

$$NADP + H_2O \xrightarrow[\text{chloroplasts}]{\text{light}} NADPH_2 + \tfrac{1}{2}O_2$$

The carriers which have been identified include chlorophylls a and b, quinones, cytochromes b and f, Fe-S centres plastocyanin, reductase enzymes and ferredoxin.

ATP formation accompanies the transfer of electrons from the quinones to cytochrome f and involves side reactions which are obligatorily coupled to the electron transfer process. Thus $NADPH_2$ and ATP formation occur in the process of non-cyclic photophosphorylation where electrons are removed from H_2O and donated to NADP with the coupled formation of ATP. This overall reaction can be expressed as:

$$NADP + H_2O + 2ADP + 2P_i \xrightarrow[\text{chloroplasts}]{\text{light (2e}^-)} NADPH_2 + 2ATP + \tfrac{1}{2}O_2$$

This equation implies that each H_2O is split in the chloroplast membrane under the influence of light to give off $\tfrac{1}{2}O_2$ molecule (an atom of oxygen) and that the two electrons so freed are then transferred to NADP, along with the H^+'s (protons) from the H_2O, to produce the strong reducing agent, $NADPH_2$. Two molecules of ATP can be simultaneously formed from two ADP and two P_i (inorganic phosphate) so that energy is stored in the form of this high energy compound.

The $NADPH_2$ and ATP are the 'assimilatory power' required to reduce CO_2 to carbohydrate in the dark phase, which will be discussed in the next chapter. Thus 'assimilatory power' represents the initial products of the conversion of light energy into chemical energy.

A diagrammatic representation of the electron flow pattern in non-cyclic phosphorylation is given in Fig. 5–2 (a). This formulation is derived from the elegant hypothesis of Hill and Bendall in 1960. The scale on the left shows very clearly the potential of all the electron carriers in the chain and implies that the sequence of electron flow depends to a large extent on their potential. This a very neat view of electron transport and all the evidence to date indicates that it is most probably correct. The orientation of the

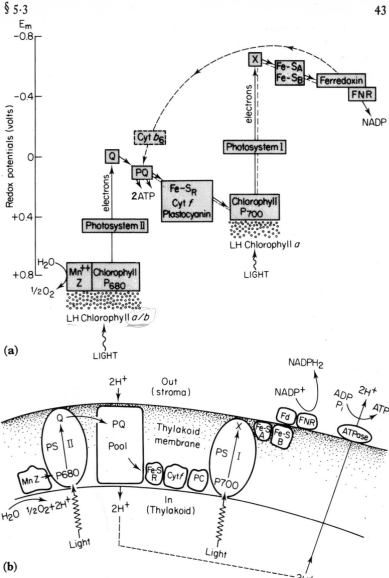

Fig. 5–2 (a) Electron transport scheme in chloroplasts. Bold line: Non cyclic electron transport from water to NADPH₂; dashed lines: cyclic electron transport through PSI. Z, Primary e⁻ donor of PSII; Q, primary e⁻ acceptor of PSII; PQ, plastoquinone; Fe-S$_R$, Rieske protein; LH, light-harvesting; X, primary e⁻ acceptor of PSI; Fe-S$_A$ and Fe-S$_B$, iron-sulphur centres; FNR, ferredoxin-NADP reductase. **(b)** Proposed relative orientation of the components of the electron transport chain across the thylakoid membrane. (After BOLTON and HALL, 1979).

electron carriers within the membrane itself is presently envisaged as in Fig. 5-2(b).

What is immediately apparent is that two different light reactions are required to raise the electrons from the level of H_2O (+0.82 V) to the level of $NADPH_2$ (-0.34 V) – these are designated Photosystems I and II. Each of the systems has a different type of chlorophyll as the main light-absorbing pigment. System I requires chlorophyll a with an absorption maximum of 660 nm (bluish-green in colour) while Photosystem II also requires the closely related chlorophyll b (see Chapter 3) which has its main absorption peak at 643 nm (yellowish-green in colour). Chlorophyll a occurs in all plants and algae, while chlorophyll b, an accessory pigment, is present in plants and green algae – blue-green algae may contain phycocyanin and allophycocyanins and red algae phycoerythrin as their accessory pigments for Photosystem II.

The first evidence for the possible involvement of two light reactions in photosynthesis came from the work of Emerson and co-workers from 1943 on (see also Chapter 4). They showed that the reduction of one molecule of CO_2 to carbohydrate required light of two different wavelengths if one of the lights was *only* absorbed by chlorophyll a, e.g. light of wavelength greater than 680 nm. The wavelength of the second light required to give efficient photosynthesis was found to correspond to that of chlorophyll b in higher plants and to the other accessory pigments in lower plants.

Further evidence for the requirement of light of two different wavelengths in non-cyclic electron flow from H_2O to NADP came from experiments which measured the changes in oxidation and reduction state of the cytochromes in algal chloroplasts. In Fig. 5-2 it is seen that cytochrome f is an electron carrier intermediate between Photosystems I and II. Using very sensitive spectrophotometric techniques (see Chapter 4) it is possible to measure the redox state of the cytochrome f, due to its specific absorption peaks at 422 and 550 nm, under illumination of light of different wavelengths being absorbed by the two Photosystems. The experimental results are shown in Fig. 5-3. Cytochrome in the chloroplast is naturally in the reduced state in the dark and thus if the algae are illuminated with light of 680 nm wavelength, which is mainly absorbed by Photosystem I, i.e. chlorophyll a in the red alga, *Porphyridium*, electrons are removed from cytochrome f and donated to ferredoxin and thence to NADP – thus the cytochrome f becomes oxidized. Then if light at 562 nm (absorbed by Photosystem II, phycoerythrin in *Porphyridium*, but also to some extent by Photosystem I) is applied to the chloroplast the cytochrome f becomes reduced since it accepts electrons from Photosystem II. Note that complete reduction is not achieved since Photosystem I is still functioning to a limited extent in removing the electrons from the cytochrome f. In the dark the cytochrome f returns to its normal reduced state seen at the start of the experiment. This type of experiment was initiated by Duysens in 1961.

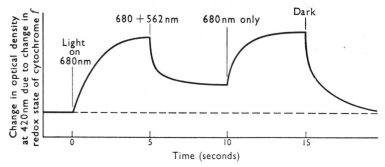

Fig. 5-3 Oxidation and reduction of cytochrome *f* in the red alga *Porphyridium*. Increase in OD_{420} is due to oxidation and decrease in OD_{420} is due to reduction of cytochrome *f*. Light of 680 nm is absorbed by Photosystem I (chlorophyll *a*) and at 562 nm is absorbed by Photosystem II (phycoerythrin). (After Duysens and Amesz, 1962, *Biochim. Biophys Acta*, **64**, 243.)

Similar experiments have recently been done to show the reduction and oxidation of cytochrome *b* in the chloroplast. This type of experiment is of great importance in localizing the site of electron carriers in the chain. The technical difficulties are great as one must be able to measure specific changes in one carrier at a time – a difficult but very rewarding technique.

Many different compounds have been isolated from chloroplasts (see Chapter 3) but the electron carriers shown in the scheme for non-cyclic photophosphorylation (Fig. 5–2) are those for which a role has been assigned so far. These compounds have been isolated and characterized chemically – quite a lot is known about ferredoxin and plastocyanin but relatively less about cytochromes b_6 and *f*. Experiments have been devised which enable one to remove a specific carrier from the chloroplasts, e.g. by washing with water or low concentrations of detergents, organic solvent extraction or by mild sonication; a certain electron transfer reaction is thus diminished and then the re-addition of the extracted compound in its purified form restores the electron-carrying ability of the chloroplast membranes. This type of experiment has been successfully accomplished with plastoquinone, plastocyanin, ferredoxin, and the enzyme (a flavoprotein) acting between ferredoxin and NADP. These types of experiments add further evidence to that accumulated in other investigations on the relative position of the electron carriers.

In the last decade low temperature (4 to 78 K) electron paramagnetic resonance spectroscopy has been applied successfully to detect a number of paramagnetic species as components of the chloroplast electron transport chain. This technique coupled with measurements of redox potentials under anaerobic conditions, has enabled investigators to assign locations for components such as the Fe–S centre R (R, Reiske), Fe–S centres A and B, and centre X (X has not been definitely characterized but may be an

Fe-quinone) in the chloroplast electron transfer chain. None of these components has been isolated from the membrane. The exact electron transfer sequence between X and ferredoxin is still not definitely established. Also there may be another component between P_{700} and X.

Elegant experiments by Levine, Bishop and others using genetic mutants of the green alga *Chlamydomonas* have also helped consolidate the electron flow sequence of Fig. 5–2. Specific mutants of the alga have been obtained which are deficient in certain parts of the electron chain, e.g. blocking electron flow between plastoquinone and cytochrome *b* or between plastocyanin and cytochrome *f*. With a knowledge of the exact site of the blockage experiments can be designed in which electrons are added and subtracted at various parts of the chain. This can be quite easily accomplished by using different types of dyes and is also used successfully in the

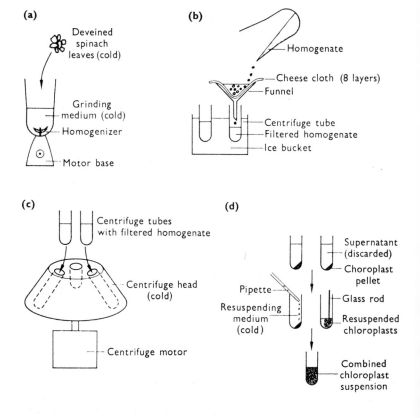

Fig. 5–4   Experiment to demonstrate photosynthetic O_2 evolution and NADPH$_2$ and ATP formation by isolated spinach chloroplasts.

(e)

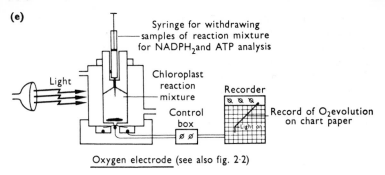

Oxygen electrode (see also fig. 2·2)

(f)

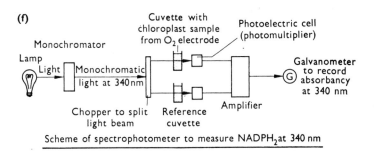

Scheme of spectrophotometer to measure NADPH₂ at 340 nm

(g)

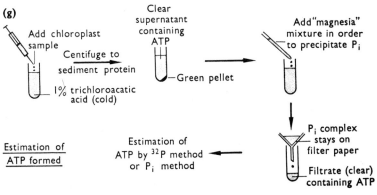

Fig. 5–4 (continued)

spectrophotometric experiments mentioned earlier. These genetic and bio-chemical experiments which are similar in principle to those used with the mould *Neurospora* and the bacterium *Escherichia coli*, have given con-firmatory evidence for the non-cyclic electron flow sequences and should prove valuable in our further elucidation of photophosphorylation and CO_2 fixation.

The use of chemical inhibitors of specific biochemical reactions is a classical and fruitful approach to understanding biochemical mechanisms. Non-cyclic photophosphorylation is no exception to the advantageous use of specific inhibitors. A number of different inhibitors have been found which block specific parts of the chain, e.g. *DCMU* [3- (3,4-dichlorophenyl) - 1, 1-dimethyl urea] a herbicide which blocks oxygen evolution; *antimycin A*, an antibiotic, which prevents the reduction of cytochrome f; and *DSPD* [Disalicylidene propane diamine] which blocks ferredoxin-catalyzed electron flow.

Lastly we must discuss the ATP formation which is coupled to the non-cyclic electron flow. In 1958 Arnon, Whatley and Allen demonstrated the obligatory coupling of ATP formation to the reduction of NADP and showed that the rate of electron flow to NADP was dependent on the presence of ADP and P_i (which is required to form ATP). How many ATP molecules are formed per $NADPH_2$ produced is an important question because of the amount of ATP required to fix CO_2 in the dark phase - 2 $NADPH_2$'s and 3 ATP's are required per CO_2 molecule reduced to the level of carbohydrate. The present evidence suggests between 1 and 2 ATP per $NADPH_2$. The mechanism of coupling between the electron transfer chain and the formation of ATP itself is thought to involve high-energy states and so-called 'coupling factors' which are proteins associated with the chloroplast membrane.

Mitchell has proposed a 'chemiosmotic' theory to account for ATP formation in membranes. This hypothesis envisages ATP synthesis occurring as the result of the recombination of positive and negative charges across the grana membrane - the initial separation of the charges being brought about by the transfer of electrons along the electron carrier chain. This idea requires that the membranes have a definite orientation and ability to bring about this charge separation (see Fig. 5-2b); it is attractive since the chloroplast membranes are similar to other membranes involved in ATP synthesis and breakdown, e.g. mitochondria, and the theory is proposed to work for all types of membranes (see TRIBE and WHITTAKER, 1981).

During the isolation and preparation of chloroplasts from the whole leaf the ATP formation factors are easily destroyed so that greater care must be taken in performing phosphorylation experiments than those in which only electron transport is measured. In Fig. 5-4 a sequence of diagrams shows how one isolates chloroplasts and measures the O_2 evolution, $NADPH_2$ formation and ATP formation associated with non-cyclic photophosphorylation (see also Chapter 3).

5.4 Cyclic electron transport and phosphorylation

In this process, which requires light and chloroplasts, the only net product is ATP. The reaction was discovered in 1954 by Arnon, Allen and

Whatley using isolated spinach chloroplasts and by Frenkel using chromatophores isolated from photosynthetic bacteria. It may be very simply
represented by the following equation:

$$ADP + P_i \xrightarrow[\text{chloroplasts}]{\text{light}} ATP$$

Only a cyclic electron flow involving Photosystem I is required in order
to produce ATP. Figure 5–2 (a) suggests how this may occur. Under the
influence of an input of light an electron is removed from P_{700} in its
excited state and donated to Fe–S centres and subsequently to ferredoxin
which becomes reduced. The reduced ferredoxin then, instead of trans-
ferring its electron to NADP as in the case of non-cyclic electron flow,
donates its electron to cytochrome b and thence through the electron
transport chain back to P_{700}. Thus the electron undergoes a cyclic flow
and the only measurable product is ATP which is formed by a coupling
mechanism, probably similar to that involved in non-cyclic photophosphorylation even though the electron carriers may not be identical. The
number of molecules of ATP formed per electron transferred is so far
undetermined because of the difficulty in measuring the number of
electrons cycling around the chain in a given time – the number of ATP's
formed in a given time is, on the other hand, relatively simple to measure.

We can thus see that ferredoxin may play a central role in photosynthesis.
It can donate electrons in a non-cyclic system to NADP in order to produce the strong reducing power in the form of $NADPH_2$ needed for CO_2
reduction; or it can donate electrons back into the electron transfer chain
in a cyclic system resulting only in the formation of ATP. This ATP can
be used for CO_2 fixation or for other reactions which only require ATP as
their energy source, e.g. protein synthesis and the conversion of glucose to
starch, both of which occur in the chloroplast. The physiological controlling
mechanism associated with ferredoxin and its very low reducing potential
of -0.43 V (equivalent that of H_2 gas) have stimulated much research into
the physicochemical and biochemical properties of this unique protein. A
model for the active centre of ferredoxin is given in Fig. 5–5.

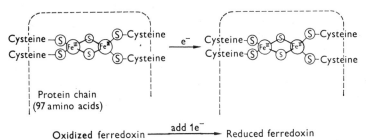

Fig. 5–5   A model for the active centre of plant ferredoxins showing the
oxidized and reduced forms of the protein. (After Rao, Cammack, Hall and
Johnson, 1971, *Biochem, J.*, **122**, 257.)

In the chloroplast ferredoxin is probably the physiological carrier (or co-factor) involved in cyclic photophosphorylation. However, experimentally we can replace ferredoxin by vitamin K_3, FMN or any of a number of dyes. The cyclic electron flow sequence may not be exactly the same as that for ferredoxin but again the only net product is ATP.

It is also possible for the reduced ferredoxin to be oxidized by oxygen (rather than NADP) in the chloroplast under certain conditions. This non-cyclic electron transport from water to O_2 is designated as pseudo-cyclic electron transport. Such pseudocyclic electron transport results in O_2 uptake and formation of H_2O_2 ('Mehler reaction') which must be dissipated from the chloroplast since it can be toxic. The formation of ATP accompanying pseudocyclic electron transport is referred to as *pseudocyclic photophosphorylation*. The possible role of such an oxygen-catalysed pseudocyclic electron transport and phosphorylation in the regulation of CO_2 fixation and the protection of the chloroplast from damage by excess light is currently being investigated.

5.5 Structure–function relationships

The light phase of the overall CO_2 fixation to the level of carbohydrate has been shown to occur in the grana lamellae (or thylakoids) of the chloroplast while the dark phase occurs in the stroma of the chloroplast. Arnon and co-workers demonstrated this in 1958 by physically separating the light and dark phase (see Table 4). Chloroplasts were illuminated in the absence of CO_2 and allowed to form large amounts of $NADPH_2$ and ATP with the concomitant O_2 evolution from the non-cyclic electron flow. The chloroplasts were then broken and the stroma separated from the grana lamellae which were discarded. In the dark, radioactive CO_2 was supplied and the enzymes in the stroma then proceeded to assimilate the CO_2 to produce the same carbohydrates that whole chloroplasts and intact green leaves manufacture.

Table 4 Carbon dioxide fixation in the dark and light by chloroplast systems, i.e. stroma (yellowish matrix) and grana (chlorophyll-containing, green membranes) (TREBST, TSUJIMOTO and ARNON, 1958, *Nature*, **182**, 351).

	$^{14}CO_2$ *fixed* (*counts per minute*)
Stroma (dark)	4 000
Stroma (dark) + grana (light)	96 000
Stroma (dark) + ATP	43 000
Stroma (dark) + $NADPH_2$ + ATP	97 000

Note the equivalence of grana (light) and $NADPH_2$ + ATP, i.e. assimilatory power.

These experiments very neatly showed that all the electron carriers and enzymes required for the light-induced $NADPH_2$ and ATP formation via cyclic and non-cyclic electron flow are associated with the chloroplast membranes (thylakoids). The enzymes for CO_2 fixation itself occur in the yellowish-coloured, amorphous stroma of the chloroplast. The task for the biochemists and electron microscopists is to localize the electron carriers in the membranes more exactly (see Figs 5–2 (b) and 8–1).

6 Carbon Dioxide Fixation

In the previous chapter we have seen that $NADPH_2$ and ATP are produced in the light phase of photosynthesis. The fixation of CO_2 then takes place in the dark phase using the 'assimilatory power' of $NADPH_2$ and ATP. In this chapter we shall examine in some detail the reactions involved in the reduction of CO_2 to the level of carbohydrate since the reaction mechanisms and experimental techniques, so clearly worked out by Calvin and his co-workers from 1946 on, are some of the most important in modern biology. For his work in elucidating the path of carbon in photosynthesis Calvin received the Nobel Prize for Chemistry in 1961.

6.1 Experimental techniques

When the long-lived isotope of carbon, ^{14}C, became available in 1945 its use, coupled with two-dimensional paper chromatography developed a few years earlier, enabled experiments to be devised to investigate the pathway of photosynthetic [14]CO_2 fixation. The unicellular green algae *Chlorella* and *Scenedesmus* were used in the experiments because of their biochemical similarity to higher green plants and because they could be grown under uniform conditions and subsequently very quickly killed in the short-time experiments used.

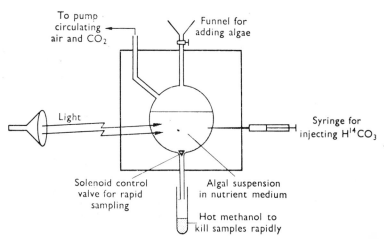

Fig. 6–1 Schematic representation of apparatus for studying [14]CO_2 fixation in photosynthesizing algae. (Diagrammatic; from CALVIN and BASSHAM, 1962.)

Three main types of experiments were performed to obtain the evidence required to postulate the detailed reactions of the cycle:

(a) Exposure of the photosynthesizing algae to $^{14}CO_2$ for different lengths of time. At the shortest times only the initial products will be radioactive. In this way phosphoglyceric acid (PGA) was identified as the primary carboxylation product; end-products such as sucrose became radioactive much more slowly.

(b) Determination of the position of radioactivity within the labelled compounds. In this way the details of the interconversions of sugar phosphate to regenerate the specific sugar phosphate which accepts the $^{14}CO_2$ molecule and the mechanism of synthesis of sugars and other compounds were worked out.

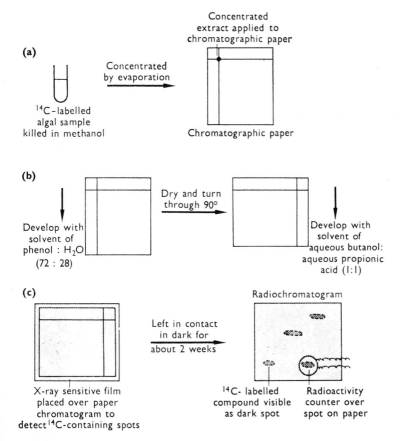

Fig. 6–2   Detection of the products of $^{14}CO_2$ fixation by algae after brief periods of illumination by the use of paper chromatography and autoradiography.

(c) Alteration of the external conditions, e.g. changing from light to dark, or changing from high to very low CO_2 concentrations, to see whether the cycle intermediates behave in a predictable manner.

The techniques employed are pictured in Figs 6–1, 6–2 and 6–3. Figure 6–1 is a diagram of the apparatus used for obtaining extracts of algae which have been photosynthesizing $^{14}CO_2$. The algae are suspended in a nutrient medium through which air and CO_2 are bubbled with the pH of the whole suspension maintained constant. The control valve allows rapid removal of samples into methanol which immediately stops all reactions from proceeding further. Figure 6–2 illustrates the further handling of the inactivated algae. The killed algal extract is concentrated by vacuum and then applied directly to a chromatogram paper which is developed with different solvents in two directions at right angles to each other. The radioactivity of the compounds which are separated by this two-dimensional chromatography is measured—their location being known from radioautograms of the kind shown in Fig. 6–3. The two radioautograms in Fig. 6–3

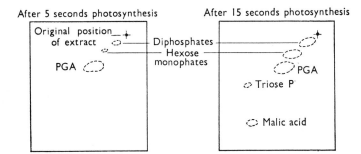

Fig. 6–3   Radioautograms of the photosynthetic products from $^{14}CO_2$ added to illuminate algae for short periods of time.

show the compounds which contain ^{14}C in extracts of *Chlorella* which had been photosynthesizing for 5 and 15 seconds. It is seen that PGA, triosephosphate and sugar phosphates are formed very rapidly; sucrose, organic acids and amino acids are only formed after longer photosynthesizing times. The combination of radioactive CO_2 and two-dimensional chromatography is seen to be a very sensitive technique for detecting and quantitatively estimating the products of photosynthesis.

6.2   The photosynthetic carbon (or Calvin) cycle

The fixation of CO_2 to the level of sugar (or other compounds) can be considered to occur in four distinct phases, as is shown in Figs. 6–4 and 6–5.

I. *Carboxylation phase*   This phase is thought to consist of a reaction whereby CO_2 is added to the 5-carbon sugar, ribulose bisphosphate, to form two molecules of PGA as follows:

$$
\begin{array}{l}
CH_2OP \\
| \\
C=O \\
| \\
*CO_2 + CHOH + H_2O \xrightarrow{\text{Enzyme}} \\
| \\
CHOH \\
| \\
CH_2OP
\end{array}
\qquad
\begin{array}{ll}
CH_2OP & CH_2OP \\
| & | \\
CHOH & + CHOH \\
| & | \\
*COOH & COOH
\end{array}
$$

Ribulose bisphosphate (RuBP) $2 \times$ Phosphoglyceric acid (PGA)

This reaction is catalysed by the enzyme ribulose bisphosphate carboxylase which is also known as carboxydismutase or Fraction 1 protein.

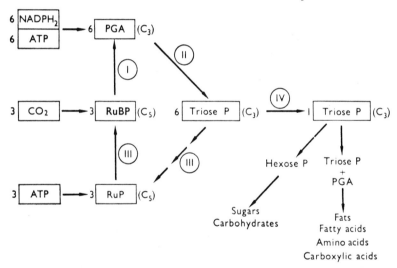

Fig. 6–4   Diagram of CO_2 fixation cycle. RuBP = ribulose bisphosphate. PGA = phosphoglyceric acid. Triose P = phosphoglyceraldehyde. RuP = ribulose-5-phosphate.

The evidence in support of this scheme is shown very clearly in Fig. 6–6. On illumination RuBP and PGA increase up to a certain level which is the so-called 'steady state' level in the photosynthesizing algae. When the light is switched off the RuBP content drops immediately (as light is needed for its synthesis) while the level of PGA rises – two molecules of PGA are formed from every molecule of RuBP which disappears. Also in Fig. 6–6 is shown the effect of changing from high to very low CO_2 concentrations. A 'steady state' level is achieved at 1% CO_2 but when the CO_2

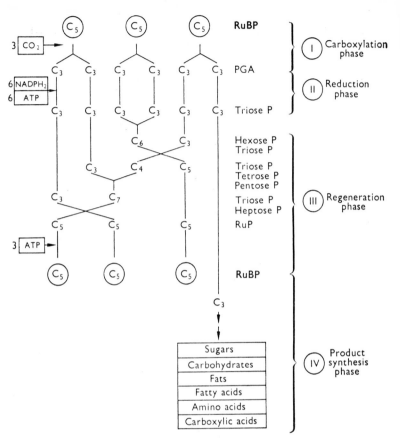

Fig. 6–5   Summary of reactions of photosynthetic CO_2 fixation. (After HALL and WHATLEY, 1967.)

level is suddenly decreased to 0.003% (with the light still on) the level of PGA drops quickly as there is insufficient CO_2 to fix; however, the level of RuBP increases since very little of it can be used to fix CO_2 into PGA, while it can still be formed in the light.

II. *Reduction phase*   PGA formed by the addition of CO_2 to RuBP is essentially an organic acid and is not at the energetic level of a sugar. In order for PGA to be convered to a 3-carbon sugar (triose P) the energy in the 'assimilatory power' of $NADPH_2$ and ATP must be used.

The reaction is in two steps, first phosphorylation, adding on a P from ATP and then reducing with $NADPH_2$, and may be summarized as follows:

$$
\begin{array}{l}
CH_2OP \\
| \\
CHOH + ATP + NADPH_2 \xrightarrow[\text{Enzymes}]{} \\
| \\
COOH
\end{array}
\qquad
\begin{array}{l}
CH_2OP \\
| \\
CHOH + ADP + P_i + NADP + H_2O \\
| \\
CHO
\end{array}
$$

Phosphoglyceric acid   Phosphoglyceraldehyde

(PGA)   (Triose P)

It is seen that the reducing power of $NADPH_2$ is used to change the acid group of PGA to an aldehyde group of the triose P; ATP is required to provide the extra energy in order to accomplish this step but the P_i of ATP

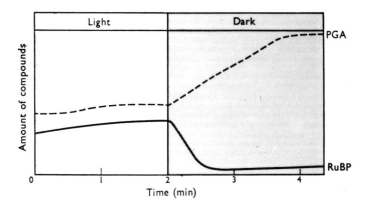

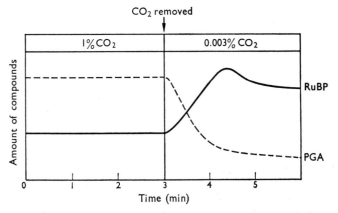

Fig. 6–6 Interconversions of RuBP and PGA, during experiments on photosynthesis. (Diagrammatic; from BASSHAM and CALVIN, in *Ruhland*, 1960.)

is not incorporated into the triose P. Both of the enzymes involved in the two steps have been shown to be present in isolated chloroplasts.

Once the CO_2 has been reduced to the level of the 3-carbon sugar, triose P, the energy-conserving part of photosynthesis has been accomplished. What is required thereafter is to regenerate the initial CO_2 acceptor molecule, i.e. ribulose bisphosphate, in order for the CO_2 fixation to continue again and again (regeneration phase) and to change the triose P to more complex sugars, carbohydrates, fats and amino acids (product synthesis phase).

III. *Regeneration phase* The RuBP is regenerated for further CO_2 fixation reactions by a complex series of reactions involving 3-, 4-, 5-, 6- and 7-carbon sugar phosphates which is depicted in summary form in Fig. 6–5. The details of the reactions are not important here but can be found in the article by BASSHAM (1962). Suffice it to say that all the reactions and the enzymes involved have been studied in some detail by various groups of research workers.

IV. *Product synthesis phase* End-products of photosynthesis are considered primarily to be sugars and carbohydrates but fats, fatty acids, amino acids and organic acids have also been shown to be synthesized in photosynthetic CO_2 fixation. Many details of these synthesis reactions are known but again they do not concern us directly. What is, however,

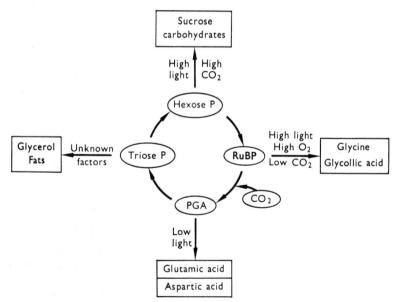

Fig. 6–7 Conditions favouring formation of secondary products in photosynthesis. (Redrawn from Hiller and Whatley, 1967, *Advancement of Science*, p. 643).

interesting is that the different end-products seem to be formed under different conditions of light intensity, CO_2 and O_2 concentration, as is depicted in Fig. 6–7. Much research is now being aimed at working out the synthetic reactions involved in the formation of these end products because an understanding of the reactions and the conditions favouring them may eventually enable us to induce plants to synthesize more or less sugars, fats or amino acids, by providing the required growth conditions.

6.3   Structure–function relationships

In order to study CO_2 fixation in isolated chloroplasts they must be isolated rather carefully in order to preserve all the components of the reactions involved. Arnon's laboratory showed in 1954 that this could be accomplished with all the CO_2 fixation products being identified. However, the rates of fixation were less than one tenth of those observed in leaves and it was not until quite recently that Walker's and Bassham's laboratories were able, by using very careful isolation media and procedures, to obtain chloroplasts capable of CO_2 fixation rates approaching those in whole leaves. Again, the products of photosynthesis were the same as those observed by Calvin's group in whole algae and by Arnon's laboratory in the early isolated chloroplasts.

These studies emphasize that structural integrity is important in understanding how subcellular organelles such as chloroplasts do actually function. The way is now clear to study many more product synthesis reactions in isolated chloroplasts, e.g. starch and sucrose synthesis. Movement of inorganic phosphate and sugar phosphates in and out of the chloroplast is strictly controlled depending on metabolic requirements. However, basically we can say that the light phase occurs in the grana lamellae or membranes and that the dark phase occurs in the stroma or soluble part of the chloroplast.

6.4   Energetics of CO_2 fixation

If we look at the overall and constituent equations of photosynthesis again, we will be able to examine the energy conserving and energy expending parts of the carbon fixation cycle.

The general equation for formation of glucose can be represented by:

$$(i) \;\; CO_2 + H_2O \rightarrow [CH_2O] + O_2 \qquad \Delta G = +48 \times 10^4 \text{ J (114 kcal)}$$

This means that 48×10^4 joules of energy are required to fix one mole of CO_2 to the level of glucose. This large positive value thus requires that a large amount of energy must be added.

We have already seen that this energy is derived from the light phase of photosynthesis and can be represented by the 'assimilatory power' of $NADPH_2$ and ATP. In order to fix one CO_2 molecule, two molecules of $NADPH_2$ and three of ATP are required (see Figs 6–4 and 6–5).

The energy present in the $NADPH_2$ and ATP can be represented as follows:

(ii) $2NADPH_2 + O_2 \rightarrow 2NADP + 2H_2O$

$$\Delta G = -44 \times 10^4 \text{ J } (-105 \text{ kcal})$$

(iii) $3ATP + H_2O \rightarrow 3ADP + 3Pi$

$$\Delta G = 9.2 \times 10^4 \text{ J } (-22 \text{ kcal})$$

This energy is sufficient to reduce one CO_2 molecule to the level of glucose with about 5×10^4 joules (13 kcal) to spare (ii + iii − i).

The constituent parts consist of

	ΔG (J)	ΔG (kcal)
(i) $CO_2 + H_2O$ $\rightarrow [CH_2O] + O_2$	$+48 \times 10^4$	$+114$
(ii) $2NADPH_2 + O_2$ $\rightarrow 2NADP + 2H_2O$	-44×10^4	-105 (2×52.5)
(iii) $3ATP$ $\rightarrow 3ADP + 3Pi$	-9.2×10^4	-22 (3×7.3)
	-5×10^4 J (excess)	-13 kcal (excess)

In summary we can present the equation as:

$$CO_2 + H_2O + 2NADPH_2 + 3ATP \longrightarrow$$
$$[CH_2O] + O_2 + 2NADP + 3ADP + 3P_i$$

Thus we see that photosynthesis is essentially a reductive process since most ($44 \times 10^4 / 53.2 \times 10^4 = 83\%$) of the energy required to fix a molecule of CO_2 is derived from the strong reducing agent, $NADPH_2$, which has a redox potential of -0.34 V. The redox potentials of sugars can be thought to be approximately -0.43 V so the ATP is required to fix the CO_2 to this lower redox value.

We know that the redox potential of $H_2O \rightarrow O_2$ is $+0.82$ V, so the overall change in redox potential is 1.25 V ($+0.82$ to -0.43 V). This can be converted into terms of energy using the equation

$$\Delta G = -nF\Delta E$$
$$= -(4)(9.64 \times 10^4)(1.25) = 48.2 \times 10^4 \text{ J } (116 \text{ kcal})$$

where n = number of electrons ($=4$ electrons per molecule of oxygen)

F = the Faraday ($=9.64 \times 10^4$ J per volt equivalent)

ΔE = difference in redox potential

It is seen that this ΔG value is very close to that for fixing one molecule of CO_2 (48×10^4 J $= 114$ kcal; equation above) and shows quite nicely the interconversion of energy in terms of joules and redox potentials.

Lastly, we can discuss the quantum efficiency of CO_2 fixation. Each quantum of red light at 680 nm contains 17.61×10^4 J of energy. Thus at least three $(48 \times 10^4/17.61 \times 10^4 = 2.7)$ quanta of 680 nm light will be required for one CO$_2$ molecule to be fixed. However, experimentally it is found that 8–10 quanta of absorbed light are required for each molecule of CO$_2$ fixed or O$_2$ evolved. From our knowledge of non-cyclic photosynthetic phosphorylation we deduce that there are two different light reactions required to reduce NADP with the electrons from H$_2$O

$$2\text{NADP} + 2\text{H}_2\text{O} \xrightarrow[\substack{\text{2 light reactions} \\ \text{chloroplasts}}]{4e^-} 2\text{NADPH}_2 + \text{O}_2$$

Thus we need at least 8 quanta (4 quanta per 4es (one O$_2$ molecule) ×2 light reactions) to reduce NADP and produce the necessary ATP at the same time.

Nevertheless photosynthetic CO$_2$ fixation itself is only about 30% efficient (2.7 quanta/8–10 quanta) as we can measure it. Taken in conjunction with an average efficiency of less than 1% for whole plants capturing and utilizing photosynthetically active sunlight (see Chapter 1) this reinforces the concept that these energy exchanges can be very wasteful processes and could be improved.

6.5   The C$_4$ pathway of CO$_2$ fixation

Many tropical grasses and plants such as sugar cane and maize are able to fix CO$_2$ initially into 4-carbon compounds like oxaloacetate [COOH–CO–CH$_2$COOH], malate [COOH–CHOH–CH$_2$COOH] and aspartate [COOH–CH$_2$–CHCOOH] in addition to CO$_2$ fixation by the Calvin C$_3$
$$\overset{|}{\text{NH}_2}$$
cycle. As already mentioned (p. 29) the leaves of these C$_4$ plants possess two types of chloroplasts, mesophyll and bundle sheath. The stomata of the C$_4$ plants are usually located such that the substomatal cavity is immediately adjacent to the mesophyll cell chloroplasts. The CO$_2$ which diffuses into the leaf through the stomata enters the mesophyll cytoplasm where it reacts with phosphoenol pyruvate [PEP] to form oxaloacetate in the presence of the enzyme PEP carboxylase.

$$\text{CO}_2 + \text{PEP} \left[\text{CH}_2 = \overset{\overset{\text{OP}^-}{|}}{\text{C}} - \text{COOH}\right] \xrightarrow[\textit{carboxylase}]{\textit{PEP}} \text{oxaloacetate}$$

There is a high concentration of PEP carboxylase in the mesophyll cells of C$_4$ species and thus CO$_2$ can be readily fixed down to low concentrations. The oxaloacetate is subsequently reduced by NADPH$_2$, formed by the normal light reactions, to malate. Radioactivity-labelling experiments using ^{14}CO$_2$ have shown that more than 90% of the radioactivity is fixed in C$_4$ acids in one second. The malate is then transported to the bundle sheath cells where it is decarboxylated to pyruvate and CO_2 and the CO_2 thus released is then used for sugar and starch production via the Calvin C_3 cycle. The malate can also function as a constituent of the Kreb's cycle or can be aminated to aspartate and form a constituent of the amino acid pool. A simplified scheme of CO_2 fixation by C_4 plants is given in Fig. 6–8.

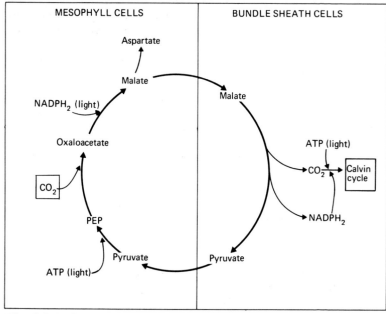

Fig. 6–8 CO_2 fixation scheme in C_4 plants.

The rate of photosynthetic CO_2 fixation by C_4 plants is not affected by ambient concentrations of O_2 (which is high) and CO_2 (low) both factors which normally enhance the photorespiration (see later) rate of C_3 plants. The water use efficiency, i.e. the ratio of the mass of CO_2 assimilated to water transpired, of C_4 plants is often twice that of C_3 species. Also salinity tolerance is a common feature of many C_4 species. All these traits allow C_4 plants to survive in dry and saline habitats. Though C_4 species are of tropical zone origin and of wide occurrence in drought zones their physiological adaptation to cool temperate zones has occurred. For example Long and Woolhouse have found many characteristics of C_4 plants in the marsh grass *Spartina townsendi* growing in the coastal areas of Britain.

6.6 Crassulacean acid metabolism

Many succulent plants growing in arid environments fix CO_2 in the dark to the C_4 acids oxaloacetic and then malic – the phenomenon was investigated extensively in the Crassulaceae and termed Crassulacean acid metabolism (CAM). CAM is widespread in the angiosperm families Agavaceae, Bromeliaceae, Cactaceae, Crassulaceae, Euphorbaceae, Liliaceae, Orchidaceae, etc. CAM plants normally close their stomata during the day to prevent water loss. Their stomata open at night. CO_2 enters the leaves and combines with PEP (a product of starch metabolism) to form oxaloacetic acid in the presence of the enzyme PEP carboxylase which is found in the cytoplasm of the leaf cells. The oxaloacetate is reduced by malic dehydrogenase to malic acid, which accumulates in the leaf vacuoles. During the day the stomata become closed, the malate is transported to the cytoplasm where it is decarboxylated by a malic enzyme to yield pyruvate and CO_2. The CO_2 thus released enters the chloroplasts where it is fixed to sugars by the photosynthetic Calvin C_3 cycle. Thus in CAM plants the fixation of CO_2 to malate at night and its decarboxylation to CO_2 and pyruvate during the day are separated in time whereas in C_4 plants the two phases are separated spatially where the primary carboxylation occurs in the mesophyll and the decarboxylation reactions in the bundle sheath. Given adequate water CAM plants may behave like C_3 species.

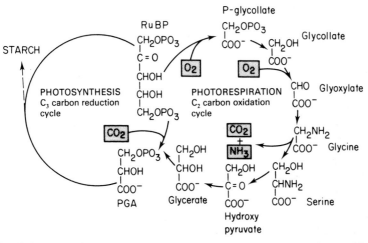

Fig. 6–9 Integrated carbon reduction and photorespiratory carbon oxidation cycle. [After Lorimer, G. H., *et al.*, 1978, *Proc. 4th Intnl. Cong. on Photosynthesis* (HALL, *et al.*, eds), p. 312.]

6.7 Photorespiration and glycollate metabolism

A very active field of current research is the study of photorespiration (the light-stimulated release of CO_2 at rapid rates by leaves) which is quite different from the 'dark' evolution of CO_2 by mitochondrial respiration in leaves. Plant species differ markedly in the rates of photorespiration; in some inefficient photosynthetic species photorespiration may be as high as 50% of net photosynthesis. Labelling experiments using the heavy oxygen isotope ^{18}O have shown that in plants with high photorespiration rates one of the initial products containing the ^{18}O label is the two carbon (C_2) acid, glycollic acid, followed in time by glycine, serine and 3-phosphoglyceric acid [PGA], a Calvin cycle intermediate. The proposed metabolic pathway of glycollic acid oxidation which leads to CO_2 formation (during the conversion of glycine to serine) is shown in Fig. 6–9. There may be additional oxidation reactions producing CO_2. How is glycollic acid formed in the chloroplasts? It was recently discovered that ribulose bisphosphate (RuBP) carboxylase, the CO_2 fixing enzyme of the Calvin cycle [p. 55] also functions as an oxygenase and catalyses the oxygenation of RuBP to 2-phosphoglycollic acid and PGA. The phosphoglycollic acid is then hydrolysed to phosphate and glycollic acid.

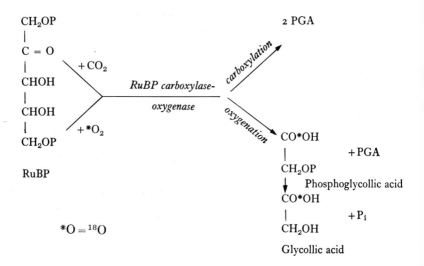

Both CO_2 and O_2 compete for RuBP at the same site of the enzyme so that high concentrations of CO_2 and low concentrations of O_2 favour carboxylation, whereas high concentrations of O_2 and low concentrations of CO_2 (as is found in the atmosphere) favour oxygenation and thus formation of phosphoglycollic acid. It has also been reported that higher temperatures favour the oxygenation reaction by the enzyme. Glycollic acid can be formed by other metabolic pathways also in the chloroplast.

The external symptoms of photorespiration are (1) the inhibition of photosynthesis by increased oxygen, (2) the existence of a high CO_2 compensation point (30–50 ppm CO_2 at 25°C in air), and (3) the variation of the CO_2 compensation point in response to variations in O_2, light and temperature.

Plants belonging to the C_4 and CAM species do not show these symp-

Table 5 General characteristics of C_3, C_4 and CAM plants (adapted from WALKER, 1979).

	C_3	C_4	CAM
1	Typically temperate species e.g. spinach, wheat, potato, tobacco, sugar beet, soya bean, sunflower.	Typically tropical or semi-tropical species e.g. maize, sugar cane, *Amaranthus*, *Sorghum*, savannah grasses. Plants adapted to high light, high temperatures and also semi-arid environments.	Typically arid zone species e.g. cacti, orchids, *Agave*, succulent plants.
2	Moderately productive. Yields of 3ot (tonnes) dry weight per hectare (2.47 acres) possible. (Sunflower is highly productive.)	Highly productive. 8ot (tonnes) per hectare for sugar cane is possible.	Usually very poorly productive. (Pineapple is highly productive)
3	Cells containing chloroplasts do not show Kranz-type anatomy and generally lack peripheral reticulum. Only one type of chloroplast.	Kranz-type anatomy and peripheral reticulum are essential features. Often have two distinct types of chloroplasts.	Lack Kranz anatomy and peripheral reticulum. Only one type of chloroplast.
4	Initial CO_2 acceptor is ribulose bisphosphate (RuBP), a 5 carbon sugar.	Initial CO_2 acceptor is phosphoenol pyruvate (PEP), a 3 carbon acid.	CO_2 acceptor is PEP in the dark and RuBP in the light.
5	Initial CO_2 fixation product is the 3-carbon acid phosphoglycerate.	Initial CO_2 fixation product is the 4-carbon acid oxaloacetate.	CO_2 fixation products are oxaloacetate in the dark and phosphoglycerate in light.
6	Only one CO_2 fixation pathway.	Two CO_2 fixation pathways separated in space.	Two CO_2 fixation pathways separated in time.
7	High rates of glycollate synthesis and photorespiration.	Low rates of glycollate synthesis; no photorespiration.	Same as C_4.
8	Low water use efficiency and salinity (ion) tolerance.	High water use efficiency and salinity tolerance.	Same as C_4.
9	Photosynthesis saturates at 1/5 full sunlight.	Do not readily photosaturate at high light.	Same as C_4.
10	High CO_2 compensation point.	Low CO_2 compensation point.	High affinity for CO_2 by night.
11	Open stomata by day.	Open stomata by day.	Open stomata by night.

toms though the kinetic properties of RuBP carboxylase isolated from these species are similar to those of this enzyme from C_3 plants or even from photosynthetic bacteria. In the C_4 plants most of the RuBP carboxylase is compartmentalized in the chloroplasts of bundle sheath cells which are not themselves in direct equilibrium with the atmospheric CO_2 or O_2. The CO_2 concentration in the bundle sheath cells may be much higher than that in the atmosphere since the CO_2 is produced *in situ* by the decarboxylation of malate which is imported from the mesophyll cells (see § 6.5) – this allows carboxylation to compete more effectively with oxygenation for the enzyme-bound RuBP. Also, any CO_2 generated by photorespiration in the C_4 species could be trapped in the chloroplasts as a result of internal recycling by the PEP carboxylase of the mesophyll cells so that the loss of CO_2 to the atmosphere is prevented. The CAM plants synthesize a high level of malate at night. Since decarboxylation of malate and CO_2 fixation by the C_3 pathway in CAM species occurs in the day when the stomata are closed [see §6.6], the internal concentration of CO_2 in these plants may well be much higher than that of the external atmosphere thus stimulating carboxylation over oxygenation. It has been shown that if C_3 plants are placed in artificial atmospheres of high CO_2 and low O_2 pressures they will behave like C_4 plants in having low photorespiratory activity.

7 Bacterial Photosynthesis

7.1 Classification of photosynthetic bacteria

Photosynthetic bacteria are typically aquatic micro-organisms inhabiting marine and freshwater environments like moist and muddy soil, stagnant ponds and lakes, sulphur springs, etc. There are three major types.

1. *Green sulphur bacteria* (Chlorobiaceae) which grow by utilizing hydrogen sulphide, or in some cases thiosulphate, as electron donor, e.g. *Chlorobium.*
2. *Purple sulphur bacteria* (Chromatiaceae) which can use hydrogen sulphide as photosynthetic electron donor, e.g. *Chromatium.*
3. *Purple non-sulphur bacteria* (Rhodospirillaceae) which are unable to use hydrogen sulphide and depend on the availability of simple organic compounds like alcohols and acids as electron donors, e.g. *Rhodomicrobium, Rhodopseudomonas* and *Rhodospirillum.*

When grown photosynthetically all three types are strict anaerobes, i.e. grow only in the complete absence of oxygen. They cannot use water as a substrate and they do not evolve oxygen during photosynthesis.

7.2 Photosynthetic pigments and apparatus

The pigment systems of photosynthetic bacteria are slightly different from those of plants and algae. The chlorophyllous pigments of bacteria are called *bacteriochlorophylls*; five classes of bacteriochlorophylls (B Chl *a*, B Chl *b*, B Chl *c*, B Chl *d* and B Chl *e*) have been characterized. These bacteriochlorophylls are very similar to chlorophylls *a* and *b* but differ in the nature of the side chains attached to the carbon atoms 2, 3, 4, 5, 7 and 10 shown for chlorophyll in Fig. 3–5. In addition a magnesium-less bacteriochlorophyll called bacteriopheophytin is found in the reaction centre of all photosynthetic bacteria. The principal carotenoids of photosynthetic bacteria are also slightly different chemically from the algal carotenoids. The nature of some of the pigments found in the photosynthetic bacteria and their growth requirements are given in Table 5. The absorption spectra of two typical bacteria are shown in Fig. 7–1.

The photosynthetic apparatus in purple and green bacteria are of morphologically different types and both types are distinct from the photosynthetic unit found in chloroplasts. The action spectra of bacteriochlorophyll fluorescence in purple bacteria indicate that light energy absorbed by carotenoids and shortwave bacteriochlorophyll bands is transferred to the longest wavelength bacteriochlorophyll (which absorb at 870 and 890 nm) before being used for photosynthesis. From measurements of substrate (carbon) assimilation and photosynthetic phosphorylation by suspensions of purple bacteria, during flashing light experiments, Clayton has estimated that the bacterial photosynthetic unit contains 30 to 50 bacteriochlorophyll molecules. When cells of purple bacteria are disrupted they release a class of subcellular particles containing all the photosynthetic pigments. These pigment-bearing particles can be isolated by the technique of differential

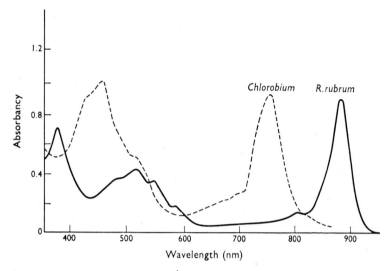

Fig. 7-1   Absorption spectra of green (*Chlorobium*) and purple (*Rhodospirillum rubrum*) photosynthetic bacteria.

centrifugation. When examined by electron microscopy after staining the particles appear like spherical bodies 30 to 100 nm in diameter and are called *chromatophores*. Each chromatophore contains several photosynthetic units. They are probably derived from the external cytoplasmic membrane by extensive invaginations (infolding) of the membrane.

Duysens showed in 1952 that the absorption spectrum of bacteriochlorophyll in purple bacteria is changed reversibly by illumination. The change corresponds mainly to a bleaching (oxidation) of the long wave absorption band of bacteriochlorophyll; at 890 nm in *R. rubrum* and *Chromatium*, and at 870 nm in *R. spheroides*.

By treating chromatophores of *R. spheroides* with detergents and then illuminating them in oxygen it is possible to destroy the light-harvesting bacteriochlorophyll molecules while still keeping the light-reacting component intact. The 870 nm absorption band of such specially prepared chromatophores is oxidized reversibly by light. This special light-reacting component in *R. spheroides* is called P_{870} and its role is similar to that of P_{700} in chloroplasts. The equivalent component in *R. rubrum* and *Chromatium* is designated P_{890}. There is one P_{870} or P_{890} molecule for about 40

Table 6 Characteristics of photosynthetic bacteria.

Group	Photosynthetic pigments	e^- donor (substrate for growth)	Growth conditions and other properties	Examples
Greens (Chlorobiaceae)	B Chl a plus B Chl c, B Chl d or B Chl e. Carotenoids Reaction centre P_{840}	H_2S $Na_2S_2O_3$ H_2	Light, autotrophic; strict anaerobes; non motile PS apparatus: *Chlorobium* vesicles and associated membranes	*Chlorobium limicola, Ch. thiosulfatophilum, Prosthecochloris aestuarii*
Purple sulphur bacteria (Chromataceae)	B Chl a or B Chl b Carotenoids Reaction centre P_{870} or P_{890}	H_2S, $Na_2S_2O_3$ H_2 Organic substrates e.g. acetate	Autotrophic and heterotrophic in light. Strict anaerobes. PS apparatus: Chromatophores	*Chromatium D Thiocapsa roseopersicina*
Purple non-sulphur bacteria (Rhodospirillaceae)	B Chl a or B Chl b Carotenoids Reaction Centre P_{870}	Organic substrates e.g. succinate, malate H_2	Heterotrophic or Autotrophic in light and anaerobic. Will grow aerobically and heterotrophically in the dark	*Rhodospirillum rubrum Rhodopseudomonas*

bacteriochlorophyll molecules which constitute the photosynthetic unit (Fig. 7–2). Difference spectroscopic studies (see Chapter 4) showed the reversible oxidation-reduction of a quinone (ubiquinone) and of a cytochrome when the light-induced changes in P_{870} were observed.

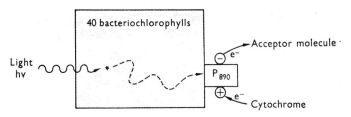

Fig. 7–2 Diagrammatic representation of the photosynthetic unit in bacteria. One photon of light reacts in a unit of 40 bacteriochlorophyll molecules containing one P_{890} reaction centre.

In recent years due to the improvements in the techniques of isolation of bacterial reaction centres and the introduction of picosecond (10^{-12}) laser flash spectrophotometry, most of the events occurring in the light reactions of bacterial photosynthesis have been elucidated. Light energy absorbed by bacteriochlorophylls and carotenoids is channelled to a reaction centre containing a few (2 or 4) specialized B Chl molecules. Charge separation occurs across the membrane at these B Chl molecules, followed by electron transport resulting in the production of ATP, $NADH_2$ or reduced ferredoxin (Fig. 7–3).

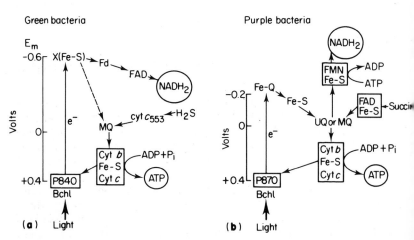

Fig. 7–3 Scheme of light-induced electron transport in photosynthetic bacteria. UQ, ubiquinone; MQ, menaquinone.

7.3 Carbon dioxide fixation

Photosynthetic bacteria do not show an Emerson enhancement effect, so it is generally considered that there is only one major photo reaction in photosynthetic bacteria. In cell-free preparations of photosynthetic purple bacteria the dominant photosynthetic reaction is cyclic photophosphorylation (production of ATP). Particles have been prepared from green bacteria which are able to catalyze photoreduction of ferredoxin (see Chapter 5) which can subsequently be used to reduce NAD to $NADH_2$ via a flavoprotein enzyme. Thus the photosynthetic bacteria can generate both ATP (energy) and $NADH_2$ (reducing power) for the fixation of CO_2 (Fig. 7–3). The Calvin–Benson cycle (Chapter 6) enzymes are shown to be present in nearly all strains of photosynthetic bacteria studied which thus have the ability to fix CO_2 via this cycle. Evans, Buchanan and Arnon have demonstrated the existence of a new route for photosynthetic CO_2 fixation in bacteria which is mediated by reduced ferredoxin leading to the synthesis of α-keto acids, e.g. pyruvate and α-ketoglutarate. Their scheme of reductive carboxylic acid cycle in photosynthetic bacteria is shown in Fig. 7–4. In this pathway of photosynthetic CO_2 fixation the ferredoxin (Fd)-dependent enzymes pyruvate synthase and α-ketoglutarate synthase catalyze the carboxylation of acetyl and succinyl-coenzyme A, respectively, as shown:

1. $\text{Acetyl CoA} + CO_2 + Fd_{red} \xrightarrow{\text{enzyme}} \text{Pyruvate} + \text{CoA} + Fd_{ox}$

2. $\text{Succinyl CoA} + CO_2 + Fd_{red} \xrightarrow{\text{enzyme}} \text{α-ketoglutaric acid} + \text{CoA} + Fd_{ox}$

The overall cycle involves the fixation of four molecules of CO_2. It is probable that in these bacteria both the reductive pentose phosphate (Calvin–Benson) pathway and the reductive carboxylic acid cycle are operating in photosynthetic CO_2 fixation.

7.4 Ecological and evolutionary significance of phototrophic bacteria

Under anaerobic (O_2–free) conditions organic matter is fermented by various micro-organisms (the chemosynthetic anaerobes) which gain their energy by a substrate-linked phosphorylation. Various metabolic end products like CO_2, H_2, ethanol and simple fatty acids are formed during this process. Such compounds would accumulate if they were not removed as nutrients by other types of microbes which are unable to use oxygen as the ultimate electron acceptor in their respiratory processes. The sulphate- and nitrate-reducing bacteria are able to consume part of the end products of fermentation of the chemosynthetic anaerobes. The phototrophic bacteria (green and purple photosynthetic bacteria) derive their energy from light and are able to metabolize most of the end products of anaerobic fermentation like alcohols, acids and hydrogen, as well as the end products

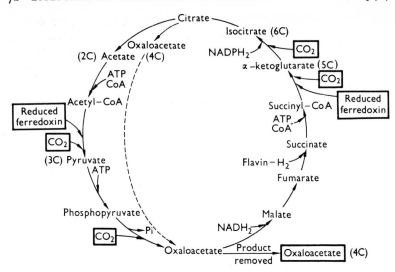

Fig. 7-4 The reductive carboxylic acid cycle of photosynthetic bacteria. Four CO_2 molecules are fixed to form one oxaloacetate molecule. (Scheme devised by Evans, Buchanan and Arnon, 1966, *Proc. Natl. Acad. Sci. U.S.*, **55**, 928.)

of sulphate and nitrate respirations such as H_2S and N_2. Thus the cell materials synthesized by the green and purple photosynthetic bacteria are future substrates for the chemosynthetic anaerobes which again produce the nutrients for the phototrophic bacteria. Thus these two types of bacteria growing in an O_2-free environment can exist together.

، Geochemical evidence suggests that the atmosphere of the early earth (prebiotic atmosphere) was oxygen-free and consisted of H_2, N_2, CH_4, NH_3, H_2S, H_2O, CO_2, etc. The ancestral photobacterium is assumed to have existed about three thousand million years ago when the earth was already two thousand million years old. In such an environment the ancestral bacterium would have utilized light to produce energy by a cyclic electron transport system. Since the oxygen of the present day atmosphere is of biological origin (the result of the photosynthetic activity of algae and plants) the green and purple bacteria represent very ancient surviving lines of organisms which at one time were the only forms able to assimilate radiant energy in the primitive, oxygen-free atmosphere.

One of the tools the biochemist uses nowadays to trace the evolutionary relationship between various taxonomic orders of organisms is a comparison of the amino acid sequence of a particular protein which is ubiquitously distributed in all these organisms. The iron–sulphur protein, ferredoxin, is found in many fermentative bacteria and in all photosynthetic bacteria and plants; cytochromes are found in all photosynthetic organisms. By comparing the amino acid composition and sequence of ferredoxins and of cytochromes of various species we are in a position to propose that the photosynthetic bacteria would occupy an intermediate period in the evolutionary history between the ancestral fermentative anaerobic bacteria and the more recently evolved algae and plants (Fig. 7–5).

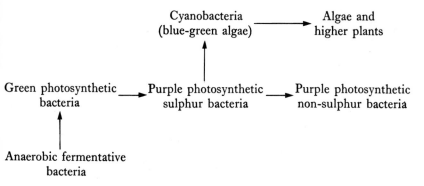

Fig. 7–5   Evolutionary scheme of photosynthesis.

Chemical Names

The names of many of the chemical substances discussed in the text have undergone changes during the last few years; the following list may be useful.

Old name	*New name*
Acetaldehyde	Ethanal
Acetic acid	Ethanoic acid (old name still acceptable)
Acetylene	Ethyne
Citric acid	2-hydroxypropane-1,2,3-tricarboxylic acid
Ethyl alcohol	Ethanol
Ethylene	Ethene
Fumaric acid	*Trans*-butanedioic acid
Glutamic acid	2-aminopentanedioic acid
Iso-citric acid	1-hydroxypropane-1,2,3-tricarboxylic acid
α-ketoglutaric acid	1-oxobutanedioic acid
Malic acid	2-hydroxybutanedioic acid
Malonic acid	propanedioic acid
Oxaloacetic acid	2-oxobutanedioic acid
Oxalosuccinic acid	1-oxopropane-1,2,3-tricarboxylic acid
Pyruvic acid	2-oxopropanoic acid
Succinic acid	Butanedioic acid

8 Research in Photosynthesis

The main areas of current research into the broad field of photosynthesis are summarized below:

Chloroplasts, protoplasts and cells

It is now relatively easy to isolate intact chloroplasts which have high rates of CO_2 fixation (Type A, complete). Thus it is possible to investigate the function of the chloroplast in relation to the role of the cytoplasm and the rest of the cell. Protoplasts are basically cells without cell walls (which have been enzymatically dissolved away) and are very useful now in studying substrate and metabolite movement into and out of the cell and the chloroplast. Individual cells (including cell walls) can also be isolated from the leaf in order to study the more complex interactions in the movement of substrates, ions, gases, etc.

Role of light in the regulation of photosynthesis

The activity of many enzymes of the CO_2 fixation pathway are modulated by light. These include NADP-glyceraldehyde-3-phosphate dehydrogenase, fructose 1,6-bisphosphatase, sedoheptulose 1,7-bisphosphatase, RuBP carboxylase, phosphoribulokinase, pyruvate phosphate dikinase and malate dehydrogenase. Several mechanisms have been proposed to explain the light activation of these enzymes such as change in stromal pH, increase in stromal Mg^{2+} concentration, change in redox state, and the meditation of protein factors and 'effectors'. The level of ascorbate in the stroma is high and the redox state of glutathione and its reductase may play a role. Two chloroplast proteins thioredoxin and ferredoxin-thioredoxin reductase have been shown as components of a ferredoxin-linked regulatory mechanism of at least four reductive pentose phosphate cycle enzymes. This is a very active area of research at present.

Starch and sucrose synthesis

The movement of metabolites between the chloroplast and the cytoplasm influences the nature and rate of photosynthesis. Sugar phosphates are exported from the chloroplast to synthesize sucrose while phosphate so released is translocated back into the chloroplast where it can be used for starch synthesis. The control of starch and sucrose synthesis in the cell and the export of sugars out of the cell and leaf are crucial to the further understanding of whole plant photosynthesis and productivity.

'Secondary' reactions

The fixation of N_2, reduction of nitrate and sulphate, synthesis of proteins, and numerous other reactions are dependent on photosynthetically produced energy sources, e.g. ATP, reduced ferredoxin and

NADPH$_2$. These types of reactions are no less important than CO$_2$ fixation and are necessary for the overall process of plant photosynthesis. The control and integration of these photosynthetic reactions is poorly understood.

Whole plant studies and bioproductivity

The efficiency of photosynthesis of the whole plant is crucial to agriculture, forestry, ecology, etc., when it comes to analysing productivity for food and fuels, and many other product uses. We need to understand how efficiency is affected by stress conditions such as temperature, water, salinity, etc., if we wish to improve plant productivity in a wide range of environmental conditions. Pollution effects are very often initially seen in their influence on the photosynthetic abilities of algae, plants and phytoplankton – again we need to understand these primary effects much more fully.

Chloroplast structure

The chloroplast envelope is now the object of much research – what special structures and enzymic properties does it have in order to perform its special functions? The molecular structure of the thylakoid membranes is being increasingly understood, i.e. protein and lipid orientations, pigment-protein complexes and their distribution, etc. (Fig. 8–1) Little is understood of the mechanism of membrane stacking into grana and the biochemical and physiological significance of this phenomenon. The relative distribution of Photosystems I and II particles in the grana and stroma membranes is also relevant in some way to the formation of grana stacks.

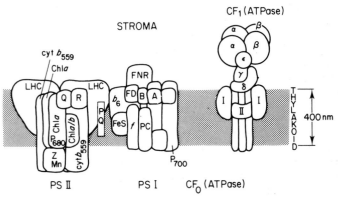

Fig. 8–1   Diagrammatic representation of the arrangement of the polypeptide components of the chloroplast membrane based on current knowledge of the electron transport chain and known orientations within the membrane. (Courtesy of Prof. D. von Wettstein, Carlsberg Laboratories, Copenhagen.)

Origin and development of chloroplasts

Did the chloroplasts of algae and higher plants (eukaryotes) evolve from the prokaryotic blue-green algae (cyanobacteria) as proposed in the endosymbiotic theory of evolution of eukaryotes? Are the *cyanelles* (symbiotic algae, see Fig. 3–5) which are found in some obligate photoautotrophs such as the *Cyanophora* sp. (a flagellated algae containing a blue-green algal symbiont) modern representatives of evolutionary links between blue-green algae and chloroplasts? There is increasing evidence supporting the close similarity between blue-green algae and the chloroplasts of eukaryotes.

Another active research area is the synthesis and assembly of chloroplast components, the synthesis of what components are directed by the chloroplast genome and what components are coded by the nuclear genome of the cytoplasm. How are the components synthesized in the cytoplasm transported to and assembled in the chloroplast membrane? How are the proplastids of dark-grown plants and algae differentiated into normal plastids and thylakoids on exposure to light?

Genetics

Knowledge of the molecular genetics of the chloroplast and its reactions is finally progressing rapidly. We know more about the chloroplast DNA, the site of synthesis and translocation of chloroplastic proteins, the regulation of protein synthesis, and so on. This should open up possibilities of manipulation of photosynthetic reactions in chloroplasts, cells, leaves and whole plants sometime in the future. The genetics of blue-green algae and photosynthetic bacteria and the role of plasmids is proving very interesting since these organisms are both considered as bacteria and thus possess many of the attributes of prokaryotic organisms as far as genetic manipulation is concerned.

The electron transport sequences in non-cyclic and cyclic photophosphorylation

The schemes given here are certainly not completely correct and must be modified as more is learnt about the reaction kinetics of individual components and hitherto unknown components. The reactions involved in O_2 evolution at Photosystem II, the primary electron acceptors, and the composition of the reaction centres are still only little known. The orientation of the individual components in the membrane is also of importance and the subject of considerable research. The role of the large pool of plastoquinones between Photosystems II and I is intriguing.

Transport of ions

Chloroplasts 'pump' various cations, e.g. Ca^{2+}, Na^+, K^+, and anions, e.g. PO_4^{3-}, Cl^-, as an important part of the movement and storage of ions in the cell. We know very little of the mechanisms involved.

In addition there is the light-induced movement of ions such as Mg^{2+} and H^+ across the thylakoid membranes which are involved in regulating CO_2 fixation and the surface charges of the membranes themselves. We still understand too little about the role of these ion movements.

Mechanism of ATP formation

The relationship of the proton pump and membrane field gradient (Mitchell's chemiosmotic hypothesis) to the formation of ATP is being actively investigated using subchloroplast membrane particles, antibodies, algal mutants, inhibitors and uncouplers, coupling factors and ATPase enzymes, electron transport control parameters in whole and broken chloroplasts, and various other techniques. The sites of ATP formation are also not clearly defined. There are two 'sites' of ATP formation coupled to proton translocation associated with Photosystems I and II. The precise stoichiometry of photophosphorylation ($NADPH_2$: ATP) is still uncertain, i.e. is the $ATP/_{2e}$-ratio 1.0, 1.33 or 2.0; the importance of an understanding of the correct ratio is relevant to the number of quanta required to fix one molecule of CO_2. Significant progress has been made on understanding the structure of the ATPase – the 5 different subunits are shown in Fig. 8–1. The properties of the subunits are becoming understood – membrane binding, inhibitor activity, proton exchange, phosphate binding, etc.

The Photosystem II oxygen-evolving reaction

This key reaction in all plant photosynthesis is still one of the major mysteries but recent work is helping to unravel the mechanism of water splitting to produce oxygen, protons and electrons. A Photosystem II protein complex has been isolated and when reincorporated into depleted membranes in lipsomes may restore O_2 evolution. The role of Mn is still unclear especially as at least two pools of Mn exist and they are difficult to detect spectroscopically. The sequence of light absorption, oxidation S states, proton ejection and oxygen release is still unclear. Progress is being made on the orientation of Photosystem II components within the membrane itself – oxygen and proton release occurs on the internal surface of the thylakoid membrane.

Mimicking photosynthesis

Photochemists and photobiologists are actively searching for synthetic systems which will split water using solar energy. Synthetic membranes made up of liposomes can be embedded with pigments, proteins and other catalysts for light-capture and electron transport. Several manganese compounds have been synthesized to investigate their suitability as water-splitting catalysts. Recently a number of semiconductor catalysts containing ruthenium and/or titanium which can liberate O_2 from water when illuminated have been prepared. Artificial systems which can photoreduce CO_2 to formate and methanol are now also being investigated. Ferredoxin is a chloroplast membrane system can be photoreduced when combined with hydrogenase or platinum to generate hydrogen from water. The advantage of these artificial systems over natural photosynthesis is that they can be optimised for maximum photosynthetic efficiency since they are not limited by the inherent physiological characteristics and requirements of whole plants.

9 Laboratory Experiments

9.1 **Relationship between chlorophyll content, starch synthesis and CO_2 fixation**

(a) Use of variegated leaves from *Pelargonium* or *Coleus*; test for starch by iodine reaction or for photosynthesis by leaf-disc technique (PAGE, p. 45).

(b) Light- and dark-grown plants; test for starch by iodine reaction after methanol extraction (BOWEN, p. 147).

(c) CO_2 requirement for starch synthesis; plants grown $\pm CO_2$ and starch detected by iodine assay (PAGE, p. 44).

(d) Bicarbonate concentration in solution and rate of photosynthesis; leaf-disc technique (PAGE, p. 45).

9.2 **Photosynthesis in *Elodea* and algae**

(a) O_2 evolution by *Elodea*; influence of light intensity and temperature; use Audus microburette to measure amounts of O_2 given off (MACHLIS and TORREY, p. 132).

(b) CO_2 uptake by *Elodea*; light requirements; use of indicator dye (bromthymol blue) to measure uptake (BOWEN, p. 141).

(c) O_2 evolution by algae, e.g. *Chlorella*, *Scenedesmus*; measured in Warburg manometer (DUNN and ARDITTI, p. 36).

9.3 **Separation of chloroplast pigments and chromatography**

(a) Pigment extraction; acetone extraction and separation into petroleum ether and methanol (MACHLIS and TORREY, p. 136).

(b) Column chromatography of leaf extracts; magnesium oxide powder developed with petroleum ether and benzene (MACHLIS and TORREY, p. 138).

(c) Column chromatography of algal extracts; tricalcium phosphate as the absorbent and developed with phosphate buffer pH 6 (DUNN and ARDITTI, p. 36).

(d) Thin-layer chromatography; silica gel on glass plate developed with petroleum ether and acetone (BOWEN, p. 129).

9.4 **Hill reaction in isolated chloroplasts**

(a) Reduction of the dye, dichlorophenol indophenol, by chloroplasts; measured with colorimeter or spectrophotometer (MACHLIS and TORREY, p. 141; BOWEN, p. 135; DUNN and ARDITTI, p. 43).

(b) Oxygen evolution in the presence of potassium ferricyanide as the electron acceptor; measured in Warburg manometer or oxygen electrode (HALL and HAWKINS, p. 151, PACKER, p. 174).

9.5   Action spectrum of CO_2 fixation in algae

E.g. *Chlorella, Scenedesmus*; measured as uptake of radioactive ^{14}C-bicarbonate in light of different wavelengths using cellophane or plastic sheets, e.g. 'Cinemoid' from Rank Strand Electric, London, WC2 (DUNN and ARDITTI, p. 38).

9.6   ATP formation by isolated chloroplasts

Measure disappearance of inorganic phosphate from reaction due to formation of ATP in the light – use of radioactive ^{32}P is not necessary (DUNN and ARDITTI, p. 43).

Reference books

BOWEN, W. R. (1969). *Experimental Cell Biology; An Elementary Laboratory Guide*, Collier Macmillan, London.

DUNN, A. and ARDITTI, J. (1968). *Experimental Physiology: Experiments in Cellular, General and Plant Physiology*. Holt, Rinehart & Winston, London.

HALL, D. O. and HAWKINS, S. E. eds. (1975). *Laboratory Manual of Cell Biology*. The English University Press.

PACKER, L. (1967). *Experiments in Cell Physiology*. Academic Press, New York.

PAGE, R. M. (1967). *Plants as Organisms: Laboratory Studies of Plant Structure and Function*. W. H. Freeman, London and San Francisco.

MACHLIS, L. and TORREY, J. G. (1959). *Plants in Action*. W. H. Freeman, San Francisco.

see also:

COOMBS, J. and HALL, D. O. (1981). *Techniques in Photosynthesis and Bio-productivity*. Pergamon Press, Oxford.

KIRBY, T. W. and CLARK, H. P. (1971). *Experimental Biology*. Oxford University Press.

KROGMANN, D. W. (1971). *Molecules, Measurement, Meanings*. W. H. Freeman, San Francisco.

SAN PIETRO, A., ed. (1980). *Methods in Enzymology*, Vol. **69**, Part C, *Photosynthesis and N_2 Fixation*. Academic Press, New York.

WITHAM, F. H., BLAYDES, D. F., and DEVLIN, R. M. (1971). *Experiments in Plant Physiology*. Van Nostrand Reinhold, New York and London.

Further Reading

Non-specialist books

ANDERSON, J. W. (1980) *Bioenergetics of Autotrophs and Heterotrophs*, Studies in Biology No. 126. Edward Arnold, London.

FOGG, G. E. (1972). *Photosynthesis* 2nd ed. Hodder & Stoughton, London.

GREGORY, R. P. F. (1976). *Biochemistry of Photosynthesis* 2nd ed. Wiley-Interscience, London.

HEATH, G. V. S. (1969). *The Physiological Aspects of Photosynthesis*. Heinemann Educational Books, London.

RABINOWITCH, E. and GOVINDJEE, (1969). *Photosynthesis*. John Wiley, New York.

TRIBE, M. and WHITTAKER, P. (1981). *Chloroplasts and Mitochondria*, Studies in Biology No. 31. Edward Arnold, London.

WHATLEY, J. M. and WHATLEY, F. R. (1980). *Light and Plant Life*, Studies in Biology No. 124. Edward Arnold, London.

Scientific American offprints

BASSHAM, J. A. (1962). *The Path of Carbon in Photosynthesis.*

BJÖRKMAN, O. and BERRY, J. (1973). *High-efficiency photosynthesis*, **229** (4), 80–93.

GOVINDJEE and GOVINDJEE, R. (1974). *The Absorption of Light in Photosynthesis.*

HINKLE, P. C. and MCCARTY, R. E. (1978). *How cells make ATP*, **238** (3), 104–23.

LEVINE, R. P. and GOODENOUGH, U. W. (1970). *The Genetic Activity of Mitochondria and Chloroplasts.*

MILLER, K. R. (1979). *The photosynthetic membrane*, **241** (4), 100–13.

Trends in Biochemical Sciences offprints

BENNETT, J. (1979). The protein that harvests sunlight, **4**, 268–71.

BUCHANAN, B. B., WOLOSIUK, R. A. and SCHURMANN, P. (1979). Thioredoxin and enzyme regulation, **4**, 93–6.

CHOLLET, R. (1977). The biochemistry of photorespiration, **2**, 155–9.

CIFFERRI, O. (1978). The chloroplast DNA mystery, **3**, 256–8.

ELLIS, R. J. (1979). The most abundant protein in the world, **4**, 241–4.

FOYER, C. H. and HALL, D. O. (1980). Oxygen metabolism in the active chloroplast, **5**, 188–91.

HATCH, M. D. (1977). C_4 pathway of photosynthesis: mechanism and physiological function, **2**, 199–202.

HEBER, U. and WALKER, D. A. (1979). The chloroplast envelope – barrier or bridge, **4**, 252–6.

MARRS, B., WALL, J. D. and GEST. H. (1977). Emergence of the biochemical genetics and molecular biology of the photosynthetic bacteria, **2**, 105–8.

More specialized books and articles

AKOYUNOGLOU, G. (ed.) (1981). *Proc. V Intl. Congress of Photosynthesis.* International Science Services, Rehovot, Israel.

BARBER, J. (ed.) (1976–1979). *Topics in Photosynthesis.* Vol. 1–3. Elsevier, Amsterdam.

BASSHAM, J. A. (1977). Increasing crop production through more controlled photosynthesis. *Science*, **197**, 630–8.

BERRY, J. and BJÖRKMAN, O. (1980). Photosynthetic response and adaptation to temperature in higher plants. *Ann. Rev. Plant Physiol.*, **31**, 491–543.

BJÖRN, L. O. (1976). *Light and Life.* Hodder and Stoughton, London.

BLANKENSHIP, R. E. and PARSON, W. W. (1979). The photochemical electron transfer reactions of photosynthetic bacteria and plants. *Ann. Rev. Biochem.*, **47**, 635–54.

BOLTON, J. R. and HALL, D. O. (1979). Photochemical conversion and storage of solar energy. *Ann. Rev. Energy*, **4**, 353–401.

BUCHANAN, B. B. (1980). Role of light in the regulation of chloroplast enzymes. *Ann. Rev. Plant Physiol.*, **31**, 341–74.

CARLSON, P. (ed.) (1980). *The Biology of Crop Productivity.* Academic Press, New York.

CHAPMAN, D. J. and RAGAN, M. A. (1980). Evolution of biochemical pathways: evidence from comparative biochemistry. *Ann. Rev. Plant Physiol.*, **31**, 639–78.

CIBA Fndn. Symposium 61 (1979). *Chlorophyll Organization and Energy Transfer in Photosynthesis.* Exerpta Medica, Amsterdam.

CLAYTON, R. K. and SISTROM, W. R. (eds.) (1978). *The photosynthetic bacteria.* Plenum Press, New York.

CLAYTON, R. K. (1980). *Photosynthesis, Physical Mechanisms and Chemical Patterns.* Cambridge University Press.

GIBBS, M. and LATZKO, E. (eds.) (1978). *Encyc. Plant Physiol.*, Vol. 5. Photosynthesis II. Regulation of photosynthetic carbon metabolism and related processes. Springer-Verlag, Berlin.

GOVINDJEE (ed.) (1975). *Bioenergetics of Photosynthesis.* Academic Press, London.

HALL, D. O., COOMBS, J. and GOODWIN, T. W. (eds.) (1978). *Proc. of IVth Int. Cong. on Photosynthesis.* The Biochemical Society, London.

HALL, D. O. (1979). Solar energy through biology – past, present and future. *Solar Energy*, **22**, 307–23.

HALLIWELL, B. (1978). The chloroplast at work. *Prog. Biophys. Mol. Biol.*, **33**, 1–54.

HILL, R. (1965). The Biochemist's green mansions: the photosynthetic electron transport chain in plants, in *Essays in Biochemistry*, Vol. I. Academic Press, London.

JENSEN, R. G. and BAHR, J. T. (1977). Ribulose 1,5-bisphosphate carboxylase-oxygenase. *Ann. Rev. Plant. Physiol.*, **28**, 379–400.

KIRK, J. T. O. and TILNEY-BASSETT, R. A. G. (1978). *The Plastids*, 2nd edn. Elsevier, Amsterdam.

LEHNINGER, A. L. (1975). *Biochemistry*, 2nd edn. Worth, New York.

MCCARTY, R. E. (1980). Photosynthetic phosphorylation by chloroplasts of higher plants. *Photochem. Photobiol. Rev.*, **5**, 1–48.

METZNER, H. (ed.) (1978). *Photosynthetic Oxygen Evolution.* Academic Press. London.

NOBEL. P. S. (1974). *Introduction to Biophysical Plant Physiology.* W. H. Freeman, San Francisco.

OSMOND, C. B. (1978). Crassulacean acid metabolism: A curiosity in context. *Ann. Rev. Plant Physiol.*, **29**, 379–414.

RAO, K. K., HALL, D. O. and CAMMACK, R. (1981). The photosynthetic apparatus, in *Biochemical Evolution*, ed. H. Gutfreund, pp. 150–202. Cambridge University Press, Cambridge.

REINERT, J. (ed.) (1980). Chloroplasts, Vol. 10 in *Results and Problems in Cell Differentiation*. Springer-Verlag, Berlin.

RUHLAND, W. (ed.). (1960). *Encyc. Plant Physiology*, Vol. 5. Springer-Verlag, Berlin.

SANADI, D. R. and VERNON, L. P. (eds.) (1978). *Current Topics in Bioenergetics*, Vols 8A and B. Photosynthesis. Academic Press, New York.

SAN PIETRO, A. (ed.) (1980). *Methods in Enzymology*. Vol. 69, Part C. Photosynthesis and Nitrogen Fixation. Academic Press. New York.

SAUER, K. (1978) Photosynthetic membranes. *Accounts Chem. Res.*, **11**, 257–64.

Science (1975). **188**, No. 4188, May 9, issue devoted to 'Food'.

SHAVIT, N. (1980). Energy transduction in chloroplasts. *Ann. Rev. Biochem.*, **49**, 111–38.

SIEGELMAN, H. W. and HIND, G. (eds.) (1978). *Photosynthetic carbon assimilation*. Plenum Press, New York.

STUMPF, P. K. and CONN, E. E. (eds.) (1980). *The Biochemistry of Plants* Vol. 1, in *The Plant Cell*, ed. N. E. Tolbert. Academic Press, London.

TREBST, A. and AVRON, M. (eds.) (1977). *Encyc. Plant Physiol.* Vol. 5. Photosynthesis I. Photosynthetic electron transport and photophosphorylation. Springer-Verlag, Berlin.

VELTHUYS, B. R. (1980). Mechanism of electron flow in Photosystem II and toward Photosystem I. *Ann. Rev. Plant Physiol.*, **31**. 545–67.

WALKER, A. (1979). *Energy, Plants and Man*. Packard Pub. Ltd., Chichester, U.K.

WOOLHOUSE, H. W. (1978). Light-gathering and carbon assimilation process in photosynthesis, their adaptive modifications and significance for agriculture. *Endeavour*, **2**, 35–46.

WORTMAN, S. (1980). World food and nutrition: the scientific and technological base. *Science*, **209**, 157–64.

ZELITCH, I. (1979). Photosynthesis and plant productivity. *Chem. and Eng. News* (U.S.A.) Feb. 5. pp. 28–48.